QUANTUM PEARLS 2

Finding Spiritual Wisdom in a STEAM World

(noun) **Quantum:** discrete quantity of energy, a share or portion

(noun) **Pearls:** person or thing of great rarity and worth; **wisdom, luck, purity, and love**. Like other precious stones and crystals, they contain powerful energy and abilities. That is why some believe pearls have the power to heal the body, mind, and soul.

(noun) **Quantum Pearls: dense bits** of **wisdom and spiritual energy** that can be found buried deep in the mundane moments of ordinary everyday life.

QUANTUM PEARLS 2

Finding Spiritual Wisdom in a STEAM World

Christopher M. Jones, Ph.D.

Published by Truth Publications, LLC
www.truth-brand.com

Quantum Pearls 2: Finding Spiritual Wisdom in a STEAM World
ISBN: 979-8-9890909-4-5

Unless otherwise noted, all scripture quotations, beginning each devotional, are taken from the New International Version (NIV). Additional translations used include the English Standard Version (ESV), New Living Translation (NLT), and the New King James Version (NKJV).

To Joseph Roger Armbrust ("Roger")
July 25, 1943 – December 14, 2025

You were a friend, mentor, and true inspiration. This devotional would not happen without you.
Rest well, brother.

Why Quantum Pearls 2

Finding Spiritual Wisdom in a STEAM World

Faith and science have long been treated as distant disciplines, parallel lines that are known to not intersect. One speaks the language of wonder and worship; the other, of equations and evidence. But what if the two are not opposites at all? What if they are both expressions of the same divine spark? What if faith isn't just felt, but discovered? What if science isn't just observed, but inspired? What if art, logic, design, and belief are all chapters of the same sacred story?

This devotional invites you to explore the world through the lens of STEAM (Science, Technology, Engineering, the Arts, and Math) blended with Scripture, spiritual reflection, and creative insight. Each day connects a core concept from these fields to the heart of God and the walk of faith.

You don't have to be an expert to take this journey. You just need to be curious.

Curious about why the universe works the way it does. Curious about what your purpose is in the midst of it. Curious about the mystery and order, the rhythm and reason, the logic and love that shape creation.

Because here's the truth: **God is the Master Designer of it all.** He is the Architect behind the math. The Coder behind the systems. The Composer behind the melody. The Storyteller behind your story. The Artist behind your vision and spiritual experience.

You'll find devotionals on everything from Newton's Laws and entropy to algorithms and artificial intelligence. And in each week, you'll also encounter a special reflection "as seen through the lens of the creative spark." These entries celebrate the power of imagination, music, storytelling, poetry, and visual beauty as essential dimensions of both science and spirit.

Because science moves forward through creativity. And creativity is, at its core, problem-solving with vision and trust in a Higher Power.

If you've ever wondered how faith fits with quantum physics, how a poem can be as holy as a lab report, or how the God of galaxies also speaks through dance, design, and digital code, then this devotional is for you. Prepare to be stretched. Prepare to be stilled. Prepare to see faith in motion and beauty in every formula.

Table of Contents

How to Use This Devotional

A Bridge Between Two Worlds

This devotional lives at the intersection of two communities that do not always speak the same language.

On one side, there are readers who love devotionals. This could include adults who may have the *Our Daily Bread* devotional booklet on their kitchen table or maybe tucked away in a car door pocket. This community is comfortable with Scripture, reflection questions, and quiet moments with God. The language of Science, however, may feel unfamiliar or even intimidating.

On the other side are readers who love science, technology, engineering, the arts, and mathematics (STEAM). This community is comfortable with data, systems, equations, and design. Faith language, on the other hand, may feel foreign, overly emotional, or disconnected from their everyday intellectual life.

This book was written for both. And especially for the few who live at the intersection.

What This Devotional Is

This is a 31-day devotional built around STEAM (Science, Technology, Engineering, the Arts, and Math). Each day takes one core concept and connects it to spiritual truth.

You will encounter terms like inertia, entropy, asymptote, binary code, feedback loops, and algorithm. When a STEAM-related term appears in bold, that is intentional. The bolded words highlight the bridge. They signal that this concept from the created world carries spiritual meaning.

The goal is not to turn you into a physicist.

The goal is not to turn you into a theologian.

The goal is to help you see that the same God who inspired Scripture also designed the systems of the universe.

How to Engage With It

This is not a start-to-finish, sit-down-in-one-weekend read. It is meant to be engaged slowly. You can:

- Read one day at a time.
- Sit with a single devotional for several days.
- Skip ahead to a concept that intrigues you.
- Revisit a week when it speaks to your current season.

Each week begins with a "Portal Entrance." That short introduction frames the theme of the week and explains how the days connect to one another.

At the end of each week, you will find a "Creative Spark." These bonus reflections explore the same themes through the lens of art, imagination, rhythm, poetry, and design. They remind us that science progresses through creativity, and creativity flows from the spirit.

You do not have to understand every scientific concept to benefit. You do not have to agree with every theological nuance to engage honestly. Curiosity is enough.

What You Do Not Have to Worry About

You do not have to:

- Master the science.
- Memorize formulas.
- Defend your faith.
- Prove anything.

If you come from a devotional background and the science feels unfamiliar, let it stretch you gently. The terms are metaphors, not exams.

If you come from a scientific background and the faith language feels new, let it challenge you thoughtfully. The Scriptures are invitations, not threats.

This book is not trying to win an argument.

It is trying to open a conversation.

Why This Matters

We live in a world where science without morality can become dangerous, and faith without intellectual rigor can become shallow. We need both.

We need engineers with conscience.

We need artists with discipline.

We need believers who are not afraid of equations.

We need scientists who are not afraid of wonder.

This devotional exists to model that integration. The same Creator who formed galaxies also formed your imagination. The same laws that govern gravity also whisper about grace.

Read with openness.

Reflect with honesty.

Create forward.

And allow this journey to reshape not just how you think, but how you live.

Week 1

PORTAL ENTRANCE

Before you build anything, you need to understand the **laws that hold it together.**

This week begins with the core principles that shape the universe. This includes concepts like inertia, entropy, energy, light, and the beauty of mathematical constants. These aren't just scientific ideas. They are **spiritual metaphors** that reveal how God designed the world and how we are meant to live within it.

Physics teaches us about force, motion, and matter. But **faith** reveals the Source behind them. The One who spoke the cosmos into being didn't just set things in motion, He set them in **meaning.**

You'll learn how God operates with the precision of a physicist, the vision of an engineer, and the compassion of a loving Creator.

And in each week, including this one, you'll also encounter something new: a bonus reflection, **as seen through the lens of the creative spark.**

Because the universe isn't just efficient, it's **beautiful.**

Because truth isn't just calculated, it's also **composed.**

And because the One who designed gravity also gave us music, poetry, rhythm, and art to **feel and gain deeper insight** what we know.

This week will stretch your mind, stir your spirit, and **ignite your imagination.** So, step in. *In the beginning... there was creativity.*

Day 1

IN THE BEGINNING, PHYSICS

Passage: Genesis 1:1

[1] *In the beginning God created the heavens and the earth.*

The first verse in the Bible reads like the preamble to every scientific equation ever written: *In the beginning.* In physics, beginnings matter. Whether you're studying the expansion of the universe, the trajectory of a rocket, or the velocity of a falling object, everything depends on initial conditions.

That's why physicists are obsessed with origins. From the Big Bang to the birth of stars, from Newton's apple to Einstein's spacetime, we search for what was first – what set everything in motion. But Genesis doesn't open with formulas or forces. It opens with **God**.

Before gravity, there was glory.

Before mass and matter, there was meaning.

Before time began ticking, there was a Creator.

In science, we try to explain how. In faith, we meet the One who said *why.* God is not threatened by our curiosity; He's the source of it. He set the laws of physics in place. Laws that reflect His character: consistent, elegant, powerful, and precise. The universe didn't come from chaos; it came from **calling forth**.

When God said, "Let there be," the cosmos obeyed. Light leapt forward. Atoms took shape. Galaxies spun into existence. And with every force He unleashed (electromagnetic, gravitational, nuclear), He also established a deeper principle: **order out of nothing.**

That same principle applies to your life. God can speak into your darkness and bring light. He can create beauty from the void. He can set your life into motion, even if you're stuck.

✦ FAITH FORMULA ✦

When God is your initial condition, your trajectory changes – no matter your past.

✦ GOD'S CONSTANT ✦

Before there was time, God was. And He still is.

✦ REFLECTION QUESTIONS ✦

What "beginning" are you currently facing? Is it a new season, new challenge, new calling? Be **specific.**

How can you invite God to speak into it, just like He did in Genesis 1?

What would change if you trusted God as your first principle?

✦ FAITH AND ACTION ✦

Now that you've answered these questions…

- Ask God to guide you to act on your specific beginning.
- Ask for God's insight and understanding, so you can follow His will.
- Ask to be of honest love and service to others in whatever you begin. Stress being honest, so it's not selfish (What's in it for me?), but what's in it for God.
- God knows what you need. Be open to that knowledge and respond to it throughout your day and night.
- At the end of your day and your night of Faith and Action, thank God for His guidance.
- Start a personal notebook: Write "Notes to God" about what actions you've taken for your new beginning. You'll find this practice very helpful as you move forward.

This book also contains an empty page titled *Notes to God* at the end of each day where you can **write to God. Feel free to honestly express your thoughts, feelings, and actions to grow your relationship.**

✦ NOTES TO GOD ✦

Day 2

ENTROPY AND ORDER

Passage: 1 Corinthians 14:33

[33]For God is not a God of disorder but of peace...

If you leave a room alone long enough, it'll get messy. Not because anyone trashed it, just because that's what happens over time. Clothes pile up, dust settles, things get out of place. That's entropy.

In physics, **entropy** is the measure of disorder in a system. And according to the second law of thermodynamics, entropy always increases over time unless **energy** enters to bring things back into order. That's true in science, and it's true in our everyday lives.

Our relationships drift.

Our priorities scatter.

Our spiritual life starts to feel like a junk drawer, full but not functioning.

Unless... we invite **energy** into the system.

God is that energy. He is not a God of disorder but of peace. He steps into the chaos of our lives and says, "Let's bring this back together." The same God who hovered over the waters in Genesis 1, bringing form and structure to the formless void, still hovers over **your** life, ready to restore what's unraveling.

But here's the thing: entropy doesn't scream. It creeps. You don't always notice it right away. That's why **daily habits of faith** (prayer, study, stillness, actions of service) are so vital. They don't just bring peace in the moment; they **push back against the entropy** that tries to pull you apart over time.

✦ FAITH FORMULA ✦

Without intentional energy, disorder wins. With God at the center, peace prevails.

✦ GOD'S CONSTANT ✦

Even when life unravels, God remains a steady presence as he restores order with His peace.

✦ REFLECTION QUESTIONS ✦

Where are you seeing signs of entropy? Is it in your schedule, your spirit, your relationships? Be **specific**.

____

____

What daily practice can you use to reintroduce God's peace into that space?

____

____

How can you make space this week to invite God's order into your chaos?

____

____

✦ FAITH AND ACTION ✦

- Ask God to replace your entropy with **creative energy** and pray for guidance as you activate **your new daily practices,** so His order replaces your disorder and brings peace.
- Pay attention to your actions, and to your thoughts and feelings as you act. Do you feel His peace entering your schedule, your spirit, your relationships? Even a little bit? Don't expect perfection. Your actions, your constant practice, will see this little bit of peace grow.
- At the end of the day, write in your "Notes to God" how your actions have resulted in even the smallest change, moving you toward peace.

✦ NOTES TO GOD ✦

Day 3

NEWTON'S FIRST LAW

Passage: Hebrews 12:1

1…let us throw off everything that hinders and the sin that so easily entangles, And let us run with perseverance the race marked out for us.

Sir Isaac Newton's First Law of Motion is one of the most famous in all of physics: **An object at rest stays at rest, and an object in motion stays in motion, unless acted upon by an external force.**

In other words, things don't begin to move or change direction on their own. If it's stuck, it'll stay stuck. If it's moving, it'll keep going, unless in both cases something intervenes.

You don't need a physics degree to see this in your spiritual life.

Ever felt like you were spiritually stuck? In a routine of distraction or doubt, drifting into apathy, or just feeling like you're *going through the motions*? **That's spiritual inertia.**

Sometimes we settle into a comfort zone and stay there, not because it's holy, but because it's familiar. Newton would say, "That's just how things work." But Hebrews reminds us that God calls us to throw off everything that hinders us in order to allow His power to be the *external force* that breaks inertia and sets us in motion.

Likewise, if you're already moving in the wrong direction, toward burnout, bitterness, sin, don't expect that momentum to stop on its own. It won't. Something, or someone, has to interrupt that motion and reroute it.

That's grace. That's Jesus.

He's the force that doesn't just move us, but moves us **toward** the race we were meant to run.

✦ FAITH FORMULA ✦

Without intervention, inertia wins. With Jesus, motion becomes mission.

✦ GOD'S CONSTANT ✦

God doesn't just get you moving, He gives you a direction worth moving toward.

✦ REFLECTION QUESTIONS ✦

Are you stuck in any area of your life where you need a push from God? Please explain and be **specific**.

____

____

What momentum, good or bad, are you currently riding?

____

____

How can you invite God to be the force that breaks inertia or redirects your path?

____

____

✦ FAITH AND ACTION ✦

- Feel free to ask God for a push, for momentum to break your inertia. To move you forward at home, at work, or wherever you feel stuck.
- At the end of the day, write in your Notes to God what actions you've taken in breaking your inertia, however small.

✦ NOTES TO GOD ✦

Day 4

ENERGY CANNOT BE DESTROYED

Passage: John 3:16

[16]For God so loved the world that he gave his one and only Son, that whoever believes in him shall not perish but have eternal life.

In physics, the **Law of Conservation of Energy** says that energy cannot be created or destroyed. It can only change form.

That means the energy in a lightning bolt can become light, heat, or sound, but it doesn't disappear. It transforms. The same is true for the energy in your body when you move, breathe, or think. Energy shifts. It doesn't vanish.

This law is one of the most foundational truths in physics and one of the most profound parallels to the spiritual world.

When Jesus died, it looked like the end. His body was

broken. His ministry seemed finished. But in the greatest transformation the universe has ever witnessed, His death became the spark for **eternal life**. What looked like destruction was actually **redirection**. It was God's love changing form to conquer sin, death, and the grave. His concrete message of Eternal Life.

And that same law applies to your life.

What you think has been lost may be changing form. What looks like failure may be transformation. What feels like death may be the prelude to resurrection.

God's love, like energy, is unstoppable. It doesn't run out. It doesn't vanish. And it certainly doesn't stay dead.

When you feel like you're losing strength, remember: in God's hands, nothing is wasted. He doesn't discard your pain, your past, or your purpose. He transforms them.

✦ FAITH FORMULA ✦

In God's hands, nothing is lost, only transformed for His glory.

✦ GOD'S CONSTANT ✦

His love, like energy, cannot be destroyed. It only multiplies and changes form.

✦ REFLECTION QUESTIONS ✦

Where in your life do you feel something has "ended"? Be **specific**.

____

____

Could it be that God is transforming, not discarding, that part of your story?

____

How have you seen God take something broken and bring new life from it?

____

____

✦ FAITH AND ACTION ✦

- Ask God to transform your feelings of being lost or powerless with positive actions.
- At the end of the day, write in your Notes to God the actions you've taken. What transformation do you see, however small? It will grow with practice and God's guidance.

✦ NOTES TO GOD ✦

Day 5

THE CONSTANT OF LIGHT

Passage: John 8:12

12 When Jesus spoke again to the people, he said, "I am the light of the world. Whoever follows me will never walk in darkness, but will have the light of life."

In physics, few things are more sacred than constants. They're the fixed values that hold the universe together. And one of the most famous constants? **The speed of light**.

Light travels at roughly **299,792,458 meters per second** in a vacuum. No matter where you are in the universe, that value holds. It doesn't speed up. It doesn't slow down. It just is.

Now pause.

Jesus said, *"I am the light of the world."*

He didn't say *a* light.

He said *the* light.

Just like the physical speed of light is constant (unchanging, ever-reliable) so is the **spiritual light of Christ**. No matter where you go or how far you've drifted, His light travels at full strength, always reaching, always illuminating, always present.

In fact, in Einstein's theory of relativity, the speed of light is so constant that it becomes the baseline for measuring time and space. In other words: **light defines reality**.

So, when Jesus says He is the light, He's not just offering comfort, He's offering clarity. His truth defines what's real. His presence gives meaning to our moments. His love illuminates the dark corners of our lives.

And just like in space, where darkness is not a thing in itself, it's just the **absence of light**, in our souls, darkness flees when Jesus is present.

You don't have to figure it all out. You just must walk toward the Light.

Ask any writer or artist about faith in the light, in moving toward the blank page or canvas. Through light, they study images and actions, trusting a Muse, a Higher Power, God will stir their imaginations. Then, together with that guiding energy, they create Art!

✦ FAITH FORMULA ✦

Jesus doesn't just light the path, He *is* the constant that defines reality.

✦ GOD'S CONSTANT ✦

Wherever you are, His Light reaches you unchanged, unwavering, unstoppable.

✦ REFLECTION QUESTIONS ✦

Are you walking through a season that feels dim, confusing, or heavy? Please explain and be **specific**

____

____

What would it look like to let Jesus define your reality in that space?

____

____

How can you reflect His light into someone else's darkness today?

____

____

✦ FAITH AND ACTION ✦

- Throughout your day and night, get honest with God if you feel dark or confused. **Tell Him**. Pause and ask for direction.
- At the end of the day, write in your Notes to God even the smallest actions you've taken to help spread the light of love and service. Consider what you can do tomorrow to continue that.

✦ NOTES TO GOD ✦

Day 6

THE OBSERVER EFFECT

Passage: Psalm 33:13,15 & Jeremiah 12:3a (ESV)

13 The Lord looks down from heaven; he sees all the children of man; 15 he who fashions the hearts of them all and observes all their deeds.

3 Yet you know me, Lord; you see me and test my thoughts about you.

In quantum physics, there's a strange and fascinating principle called **the observer effect**. It says that the act of observing a system changes it. Put simply: when you watch something at the quantum level, it behaves differently than when you don't.

Think about that.

Just **looking** changes things.

Scientists don't fully understand why, but they've proven it over and over. Electrons and photons, which are tiny particles, alter their behavior depending on whether they're being watched. It's like creation knows it's being seen.

And you know what? You're not so different.

There's something about being **seen**, really seen, that shifts our behavior, softens our hearts, and shapes our choices. When no one's watching, it's easy to cut corners, hide pain, or lose our sense of worth. But when we know we're being watched, not judged but *noticed*, everything changes.

The Psalmist tells us that **God sees us**. Not from a distance, not with cold detachment, but with love, intention, and personal care. He observes our deeds, yes, but more than that, **He fashions our hearts**.

His gaze isn't just watching, it's working. He sees your effort when no one else does. He notices your hidden pain. He watches over your potential, even when you've lost sight of it yourself.

And just like in the quantum world, His observation isn't passive, it's transformative.

What I said in the earlier chapter about the writer and artist: how they study images and actions. They are following what God does: observing and responding. They take those observations and transform them into Art.

✦ FAITH FORMULA ✦

Being seen by God isn't surveillance, it's sacred. His gaze shapes your life.

✦ GOD'S CONSTANT ✦

The One who sees you also forms you. You are never invisible to Him.

✦ REFLECTION QUESTIONS ✦

In what areas of your life have you felt unseen or unnoticed? Be **specific.**

____

____

How does knowing God sees you with love, not scrutiny, change your perspective?

____

____

What might shift if you lived today as someone fully observed by grace?

____

____

✦ FAITH AND ACTION ✦

- Ask God to observe you throughout your day and night, to guide you. Also, observe yourself as you act and interact with others.
- At the end of the day, write in your Notes to God what you've observed about yourself. Has His light affected your view, your perception of yourself and others? Has it changed your direction in any way?

✦ NOTES TO GOD ✦

Day 7

SANCTIFIED CURIOSITY

Passage: Proverbs 25:2

2It is the glory of God to conceal a matter; to search out a matter is the glory of kings.

Good science always starts with a question. How does this work? What's behind that pattern? Why does this keep happening?

Curiosity has driven every great breakthrough, from the apple that fell on Newton's head to the Mars rovers wandering a dusty red planet in search of clues. But the deeper you go into science, the more you realize that asking questions doesn't weaken faith, it strengthens it.

Because God is not afraid of your questions. He welcomes them. He created us to learn. And through learning our lives become loving and meaningful.

In Proverbs, we're told that **it's part of God's glory to conceal**, but it's also part of **our glory to search**. That means curiosity is not just allowed, it's *holy*. The same

God who knit the laws of the universe into being also left just enough mystery in place to invite us closer.

Why? **Because relationship grows in the search.**

God could have downloaded the entire manual of the universe in a moment, but instead, He reveals Himself progressively: through Scripture, through nature, through experience, through Spirit. He leaves space for wonder, not to frustrate us, but to **form us**.

When you ask hard questions about suffering, purpose, calling, or science itself, you're not disqualified from faith. You're stepping into the sacred process of discovery.

So, ask.

Ask boldly.

Ask prayerfully.

Ask like someone who knows the Creator delights in the search.

✦ FAITH FORMULA ✦

God hides things not to keep us away, but to draw us closer in the seeking.

✦ GOD'S CONSTANT ✦

Your questions don't scare God. They lead you into deeper intimacy with Him.

✦ REFLECTION QUESTIONS ✦

What question about God, life, or the universe has been on your heart lately? Be **specific.**

____

____

Have you felt pressure to "just believe" instead of wrestling with that question? Please explain.

____

____

How can you honor God by bringing your curiosity into conversation with Him?

____

____

✦ FAITH AND ACTION ✦

- Now, consider writing in your Notes to God at your day's *beginning*. Be honest about the questions in your heart. How can God help you answer them? What actions can he guide you with through this day?
- At the end of the day, write to Him with any answers you've discovered about yourself. If you have more questions, ask them.
- We grow our relationships with God and humans through honest questions and discussions, through prayer and meditation.
- *Feel free to write a prayer and meditation to God about your heart's questions.*

✦ NOTES TO GOD ✦

CREATIVE SPARK

(Week 1 Close)

IN THE BEGINNING, THERE WAS RHYTHM

(as seen through the lens of the creative spark)

Passage: John 1:1

[1]In the beginning was the Word, and the Word was with God and the word was God.

Before the Big Bang, before the blueprint, before gravity even had a name, there was rhythm.

John 1 tells us that, *"In the beginning was the Word."* But in the original Greek, "Word" is **Logos**, which doesn't just mean "word." It means *logic, reason, flow.* It's where we get our word "logic" from, yes, but also **logo**, as in *symbol.* **Story. Pattern. Meaning.**

Before science charted forces and constants, the universe moved to a beat. And when God spoke, it wasn't just language, it was **vibration**. It was sound. It was **poetry becoming particle.**

Creation began with **voice**, and voice carries rhythm.

Have you ever thought about that? That perhaps the first evidence of God's creativity wasn't a planet or a proton… but a **pulse**. A divine beat or rhythm.

Everything in creation moves to rhythm:

- The earth spins.
- Waves crash.
- Hearts beat.
- Breath flows in and out.

Even subatomic particles **vibrate with energy** tiny dances that echo the first command: *Let there be.*

As artists, musicians, dancers, and poets have long known, **truth isn't only understood, it's felt.** And rhythm is how we feel structure. It's the invisible pattern that moves the visible world.

When we listen deeply to music, to poetry, to silence, we tune ourselves back to the original rhythm. The heartbeat of the Creator. The flow of the Word. If your world feels chaotic this week, try listening.

Not just for answers.

But for the rhythm.

It's still playing.

And it's calling you to create.

✦ FAITH FORMULA ✦

God's voice didn't just command, it composed. The universe began in rhythm, and your spirit still moves to that beat.

✦ GOD'S CONSTANT ✦

He is not only the architect of structure, He is the composer of beauty, harmony, and flow.

✦ REFLECTION QUESTIONS ✦

Where do you feel rhythm most deeply? In music, in nature, in silence? Be **specific.**

____

____

What creative expressions help you reconnect to God's original pulse?

____

____

How might your week change if you approached it like a song instead of a spreadsheet?

____

____

✦ FAITH AND ACTION ✦

- Write a poem to God in your Notes. An honest expression of your love for life and for Him. Or questions you have in search of Him. Don't worry about being perfect in your writing. Let the rhythm flow! It's your voice speaking to your creator in writing.
- Or draw a picture, a sketch, a rhythmic expression of your personal vision of God. Of your life as you see it or wish it to be. Is it a sunflower? A free-flying dove? Make it your image!
- As with all things, let God guide you.

✦ NOTES TO GOD ✦

Week 2

ENGINEERING DESIGN AND SYSTEMS

God the Designer. You the Project. Faith the Framework.

After laying the foundations of how the universe works, this week shifts from **what is** to **how it's built.**

Engineering is all about design: how forces are managed, how parts fit together, how stress is absorbed, and how systems stay functional under pressure. It's about plans, structures, and systems working as one.

In many ways, that's the invitation of faith: not just to believe in God, but to **trust His design** for your life, even when you can't see the final blueprint.

This week, you'll explore themes like load-bearing structures, feedback loops, maintenance rhythms, and stress testing. You'll see how God's grace shows up not just in the build, but in the **wear and tear** of life. He doesn't discard broken systems, He refines them.

And alongside the systems and structures, we'll pause to reflect on **the artistry of design.**

Because **design is both science and story.** It's found in stained glass windows, the curve of a violin, the shape of a city skyline, or the arcs in an architect's sketch. Whether you're building a bridge or painting a mural, you're reflecting the image of a God who doesn't just create **functionally**, but **beautifully.**

This week, allow yourself to be open not just to correction, but to **creation.**

Not just to stress, but to **symmetry.** Not just to systems, but to **story.** *God is your Engineer… and your Artist.*

Let's turn the page.

Day 8

BLUEPRINTS AND GOD'S PLAN

Passage: Jeremiah 29:11

11For I know the plan I have for you," declares the Lord, "plans to prosper you and not to harm you, plans to give you hope and a future."

Before a building is ever raised, before a bridge ever spans a river, before a single bolt is tightened, there are blueprints. Detailed plans. Schematics. A roadmap of how each part fits together to serve the whole.

In engineering, ignoring the blueprint leads to disaster. Why? Because the blueprint knows what the beams can hold. It knows where the weight will shift. It knows what kind of load the foundation is meant to carry. Everything depends on trusting the design.

And guess what? **Your life has a blueprint.**

You weren't thrown together by accident. You are not a random assembly of passions, pain, talents, and trials.

You are intentionally wired, structurally sound, and deeply known. That's what God was saying through Jeremiah: *"I know the plans I have for you."*

Not, "you should know." Not, "figure it out on your own." It states, **"I know."**

God knows the framework. He knows what kind of pressure you can take, what gifts you carry, and what impact you're designed to have. And sometimes, just like in engineering, **you won't see the final structure until the last bolt is in place.**

But here's the beauty: engineers build with **trust in the blueprint** even before the project is complete. So can you.

Even when life looks like scaffolding and exposed beams, God's plan is still unfolding. Even when you're stuck in the structural stage, God hasn't walked away from the drawing board.

✦ FAITH FORMULA ✦

You are built with intention. Trust the design, even when you don't see the structure yet.

✦ GOD'S CONSTANT ✦

God knows the plan. Always has. Always will.

✦ REFLECTION QUESTIONS ✦

What areas of your life feel like unfinished construction zones right now? Be **specific.**

____

____

How would trusting God's blueprint shift your attitude or decisions in this season?

____

____

Who in your life needs reminding that they are part of a purposeful design?

___

___

✦ FAITH AND ACTION ✦

- Have a morning meeting with God. What constructive actions are you planning for today? **Be specific.**
- Ask God to guide you through your planning, through your constructive actions.
- What if your day doesn't go as planned? What if a constructive action doesn't work? Ask God to guide you in pausing, studying what course to take. Have faith He will guide you if you honestly ask Him to.
- At your day's end, write in your Notes to God. Report back on how the day and night went. What actions went well? What actions went awry? How did you respond to both? How can He help you respond with faith and detailed plans tomorrow?

✦ NOTES TO GOD ✦

Day 9

LOAD-BEARING STRUCTURES

Passage: 1 Peter 5:7

7Cast all your anxiety on him because he cares for you.

In any building, whether it's a house, a bridge, or a spacecraft, there are certain parts that carry more weight than others. These are called load-bearing structures. You can't remove them without compromising the whole system.

An architect knows where the weight will be. A good engineer plans for it using steel beams, reinforced concrete, tension cables. The goal is simple: build in a way that can handle the pressure.

God does the same with us.

There are seasons in life when it feels like you're holding up more than your share. Seasons when you are carrying emotional weight, family responsibility, spiritual burdens, even community pain. You may wonder, "Why me, Lord? Why this load?" But what if the answer isn't punishment, it's purposeful placement?

Maybe God built you with the internal steel to handle what others can't. Maybe He placed you in that position not to break you, but to reveal what He built into you from the beginning.

And here's the twist: even the strongest structures need a foundation. Load-bearing doesn't mean self-sustaining. That's why 1 Peter tells us to cast our cares on Him. You can be strong and still surrender. You can be foundational and still lean on *the* Cornerstone.

The weight isn't meant to crush you. It's meant to connect you deeper into God's strength and more intentionally into the lives you're meant to help hold up.

✦ FAITH FORMULA ✦

You weren't made to carry everything, but you were made to carry something. Lean into the design, not away from it.

✦ GOD'S CONSTANT ✦

God bears your weight even when you're bearing someone else's.

✦ REFLECTION QUESTIONS ✦

What emotional or spiritual weight are you carrying right now? Be **specific.**

____

____

Are you trying to bear more than what God actually designed you to carry?

____

____

How can you surrender the burden while still honoring your strength?

____

____

✦ FAITH AND ACTION ✦

- Begin your day with prayer: Honest conversation with God about your specific emotional or spiritual weight. Ask him to guide your surrender, so he can help you carry this burden. You are never alone. God's spiritual muscle will assist you if you're willing.
- Right before you take an action today, briefly ask God to help. Your request can be silent and as brief as that one honest word: "Help."
- At your day's end, write in your Notes to God. What specific actions did God help you with? Where did you avoid asking for help? Why? Where were you both able to be of love and service to someone, even in the smallest way?
- If you believe God has helped you today, thank Him.
- How can you bless someone else?

✦ NOTES TO GOD ✦

DAY 10

FEEDBACK LOOPS

Passage: Proverbs 3:11-12 (NLT)

[11]My child, don't reject the Lord's discipline, and don't be upset when he corrects you. [12]For the Lord corrects those he loves, just as a father corrects a child in whom he delights."

In engineering, one of the most critical tools for system health is the feedback loop. It's a mechanism where the system monitors its own performance and makes real-time adjustments to stay on course.

Think of your body's temperature regulation: if you get too hot, you sweat. If you get too cold, you shiver. Your body takes in information, evaluates it, and corrects course. That's a feedback loop.

Now apply that to your spiritual life.

God has built feedback into your walk with Him, not to shame you, but to shape you. Scripture. Conviction. Wise counsel. Even unanswered prayers or closed doors can be feedback, signals that God is guiding you, refining you, and keeping you from veering off course.

But here's the thing: feedback only works if you listen to it.

Too many systems fail because they ignore the data. Too many people falter because they ignore the signs. And when correction comes, they resist it, resent it, or run from it.

But Proverbs tells us: don't despise it. That discipline? That divine nudge? That unexpected roadblock? It might just be God's feedback loop working for your good.

Correction is not rejection. It's a signal of relationship. A sign that God is invested in your progress.

✦ FAITH FORMULA ✦

Feedback isn't failure. It's grace in motion, God's way of keeping you on course.

✦ GOD'S CONSTANT ✦

He disciplines because He delights in your growth and progress. His correction is proof of His love.

✦ REFLECTION QUESTIONS ✦

Where have you sensed God correcting or redirecting you lately? Be **specific.**

____

____

Are there areas where you've ignored feedback spiritually, emotionally, or relationally?

____

____

How can you tune your heart to receive correction as a sign of care, not condemnation?

____

____

✦ FAITH AND ACTION ✦

- Ask God to guide you in paying attention as you move through your day and night. To be alert to the reality of each situation. To pause so -- rather than with fearful, knee-jerk reaction -- you can respond instead with reason and love.
- Pausing and taking a deep breath is an action. In fact, it is a form of prayer. Of silently asking God, "What shall we do now?"
- Paying attention will help you see if you're ignoring feedback or teaming with God on each activity.
- At the end of the day, write to God how you did this day and night. Did you welcome teamwork with Him, or run from it? Did you accept correction or reject it? What actions went well? What changes in your thinking and behavior can help you and God move forward tomorrow? Ask for His help in changing you.

✦ NOTES TO GOD ✦

Day 11

SAFETY MARGINS

Passage: 2 Corinthians 12:9

9But he said to me, "My grace is sufficient for you, for my power is made perfect in weakness."

In engineering, one of the most important, but often invisible, elements of any system is the safety margin.

It's the extra strength built in. The hidden reserve. The unseen capacity that allows a structure to endure more than it was ever expected to face. A beam designed to hold 1,000 pounds might actually be capable of 1,500 because the engineer planned for the unexpected.

Why?

Because real life has variables.

Because loads shift.

Because weather changes.

Because people make mistakes.

A wise designer always builds in margin.

And God? He's the wisest designer of all.

When Paul begged God to remove his weakness, God didn't say "tough it out." He said, "My grace is sufficient." In other words, grace is your safety margin.

You might feel like you're at your limit, but God has already factored in the extra pressure. His grace shows up in the *more than you can handle* moments, not to erase the challenge, but to prove His power in your weakness.

Grace is not just forgiveness. It's fortification.

You don't have to be as strong as you think.

You don't have to hold it together all the time.

You just have to trust the margin He's already built in.

✦ FAITH FORMULA ✦

God always builds in more grace than the weight you're carrying requires.

✦ GOD'S CONSTANT ✦

His strength is made perfect not in your surplus, but in your stretch.

✦ REFLECTION QUESTIONS ✦

Where in your life are you feeling stretched to your limit? Be **specific.**

____

____

Have you mistaken grace for absence because it's quiet, not loud? Please explain.

____

____

How can you rest in the truth that "God's design includes the unexpected?"

____

____

✦ FAITH AND ACTION ✦

- This morning be quiet and listen. Deep breathe. Listen. Can you feel God's quiet presence? He's there. And His grace and strength are there for the asking. Now ask for them. If you don't feel them right away, that's okay. That's why we call this Faith and Action. Have faith you'll be guided and supported.
- At the end of the day, consider where you have been stretched to the limit. Did you pay attention and ask for God's grace and guidance? Or did you try to bull through with self-will? What was the result? Write it all down in your Notes to God.
- Now write what changes you and God can make together to activate His grace and strength.
- Consider thanking God for His grace and strength. Are you willing to?

✦ NOTES TO GOD ✦

Day 12

REDUNDANCY IN DESIGN

Passage: Romans 8:28

28 And we know that in all things God works for the good of those who love him, who have been called according to his purpose.

If you've ever flown on a plane, you've probably experienced the quiet peace that comes from not thinking about all the things that could go wrong. But aerospace engineers? They think about everything. Every possible failure. Every worst-case scenario. And then they do something brilliant:

They build in redundancy.

Redundancy is the intentional inclusion of backup systems. If one engine fails, there's another. If one power supply shuts down, there's a secondary. If the computer crashes, there's a fail-safe. Because when the stakes are high, a single point of failure just isn't acceptable.

God operates the same way.

He knows life can be unpredictable. People let us down. Plans collapse. Even our own strength has limits. But Romans 8:28 reminds us: "In all things, God works for the good..." That's redundancy.

If one door closes, He opens another.

If one plan fails, He's already authored the next.

If one path disappears, He's got a better one under construction.

That doesn't mean life is painless. It means it's covered.

God doesn't rely on one outcome to accomplish His purpose in your life. He works in all things, the joyful, the confusing, the mundane, the miraculous. He doesn't need everything to go right for everything to still turn out good. I didn't say "perfect"! I said good! He backs us in moving toward positive results.

When you feel like something has broken beyond repair, don't panic. The system is still online.

✦ FAITH FORMULA ✦

If it's not good yet, God's not done, because He designed your life with backups in mind.

✦ GOD'S CONSTANT ✦

He works through all things, not just the ones that go according to plan.

✦ REFLECTION QUESTIONS ✦

Have you experienced a "backup plan" from God that turned out better than what you expected? Please explain and be **specific.**

____

____

Where are you tempted to panic because one thing isn't working?

____

____

What would trusting God's redundancy look like in your current situation?

____

✦ FAITH AND ACTION ✦

- Let's start the day writing in your Notes to God. Get honest with God right now: Admit you're not perfect. Now admit that God has given you gifts to live a productive life. What are those gifts? If you don't know, tell Him so. Ask Him to show you. If you know what some of them are, ask Him to guide you.
- Hint: You probably don't know what all your gifts are yet. The backup plan. New gifts seem to be revealed through your honest relationships with others. That's a good time to ask God to help you pay attention. You may be surprised to see what arises through your Faith and Action!
- At the end of the day, write to God about your faithful actions through the day. Explain how they may not be perfect…but they show progress in your efforts at Faith and Action.

✦ NOTES TO GOD ✦

Day 13

STRESS TESTING

Passage: James 1:2-3

[2]Consider it pure joy, my brothers and sisters, whenever you face trials of many kinds, [3]because you know that the testing of your faith produces perseverance.

In engineering, a stress test is exactly what it sounds like: you deliberately apply pressure to a material, structure, or system to see how much it can handle. You're not trying to destroy it, you're trying to **reveal its limits** and **increase its reliability.**

Bridges are stress tested. Rockets are stress tested. Even phone screens are dropped on purpose to see if they'll crack under pressure.

Why?

Because there's no such thing as *untested strength.*

And guess what? Faith works the same way.

James tells us that trials, those stress points in our lives, are not just obstacles. They're opportunities. They expose what's weak, reveal what's strong, and build something more durable than comfort ever could: **perseverance.**

No one enjoys pressure. But it has purpose.

You're not being punished, you're being proven.

Stress doesn't mean the system is broken. Sometimes it means you're ready to level up. The testing reveals where you need reinforcement, where you're more resilient than you thought, and where God is inviting you to depend more fully on Him.

And just like in engineering, the test isn't the final chapter, it's part of the certification process. The proof that you can be trusted with more.

✦ FAITH FORMULA ✦

Pressure reveals purpose. What's tested can be trusted and transformed.

✦ GOD'S CONSTANT ✦

He doesn't waste trials. He uses them to build endurance and deepen your foundation.

✦ REFLECTION QUESTIONS ✦

What current pressure point in your life could actually be a spiritual stress test? Be **specific.**

____

____

Are you resisting it or receiving it as part of God's shaping?

____

____

How has God used past trials to make your faith stronger and more stable?

____

____

✦ FAITH AND ACTION ✦

- Consider your plans for today…Where do you foresee stress occurring at home? At work? Where do you feel failure might occur? **Be specific**.
- Ask God to help you see the reality: We don't live alone in this life. Where can you and God handle the stress? Where will you need to ask for help from others? Pray for the willingness to ask for help. It's a part of God's gift in being human.
- Also ask for guidance in being of help to others. To relieve their stress where you can.
- At the end of the day, write in your Notes to God how you think you did today and tonight. How, together with God, you handled stress. Not perfectly, but faithfully. That's our key.

✦ NOTES TO GOD ✦

Day 14

THE IMPORTANCE OF MAINTENANCE

Passage: Psalm 119:105

105 Your word is a Lamp for my feet, a light on my path.

Engineers know something that too many people forget: **it's not enough to build something great, you have to maintain it.**

You can have the most advanced aircraft, the strongest bridge, or the most efficient engine, but without regular maintenance, it will fail. It's not a matter of if, but *when.*

Maintenance doesn't mean something is broken. It means you're taking care of it, so it **doesn't** break.

Now let's bring that home.

Your spiritual life is no different. You can be well-designed, deeply called, brilliantly gifted, but without regular maintenance, things drift. Wear shows up. Systems clog. And what once ran smoothly starts to grind.

Psalm 119 reminds us that God's Word is like a lamp, it lights the path forward. But lamps need oil. Engines need check-ups. Systems need resets. And souls? Souls need Scripture, prayer, stillness, worship, and community.

Not because we're weak.

But because we were never designed to run without regular connection to the Source.

Don't wait until burnout.

Don't wait until breakdown.

Create a rhythm now.

Engineers use maintenance schedules. You can too.

Maybe it's daily quiet time, weekly Sabbath, monthly reflection, or seasonal retreats. Whatever the rhythm, it's not about religious duty; it's about sustaining alignment with the God who powers your entire being.

✦ FAITH FORMULA ✦

You're not failing, you're just due for maintenance. Build a rhythm before burnout hampers or harms you.

✦ GOD'S CONSTANT ✦

He doesn't just start your engine. He walks with you mile after mile, day after day.

✦ REFLECTION QUESTIONS ✦

What maintenance practices help keep your spiritual system aligned and healthy? Be **specific.**

___

___

Where have you ignored signs of wear emotionally, mentally, or spiritually?

___

___

What small rhythms can you implement this week to stay connected to the Source?

___

___

✦ FAITH AND ACTION ✦

- Ask Him to provide it to you, to guide you in using it through your day and night, in the rhythm of Faith and Action.
- Pay attention to how his grace and energy help your thoughts, feelings, and actions.
- Pay attention especially when you feel doubt or weakness set in. Pause. Request a spiritual refill. It's always there for the honest asking.
- At the end of the day, write how you maintained your relationship with God by activating His grace and energy. Be specific.
- We may be surprised how God has helped us in ways we haven't recognized until we write and reflect on them.

✦ NOTES TO GOD ✦

CREATIVE SPARK

(Week 2 Close)

THE SACRED SHAPE OF THINGS

(as seen through the lens of the creative spark)

Passage: Ephesians 2:10 (ESV)

10 For we are God's workmanship, created in Christ Jesus to do good works, which God prepared beforehand, that we should walk in them.

The word "workmanship" in Ephesians 2:10 comes from the Greek word *poiēma.* It's where we get our word for **poem.**

You are God's poem.

His design.

His art.

And art, like engineering, is a form of problem solving. Whether you're painting a canvas or designing a suspension bridge, you're arranging elements into a **functional and beautiful whole.** That's what God does with your life.

We often think of blueprints as cold, calculated plans. But true design is never just technical, it's spiritual. The most enduring structures on earth, from the pyramids to cathedrals, weren't just engineered… **they were imagined.**

Design is what happens when structure meets soul. It's in the flow of a well-written line of code. The curve of a pottery wheel. The echo of a cathedral arch. The layout of a city that honors community.

Even creation itself, **from DNA to galaxy spirals**, follows divine patterns of proportion and beauty. The golden ratio, Fibonacci sequences, and sacred geometry remind us that the **Creator is also the Composer.**

And you? You're not a rough draft.

You're not accidental.

You are **intentional art** with a structural purpose.

The next time you create, be it a spreadsheet, a stanza, a system, or a sketch, remember: You're reflecting the God who designed the soul and wired it for wonder.

✦ FAITH FORMULA ✦

Design isn't just functionality, it's a form of worship. What you build reflects what you believe.

✦ GOD'S CONSTANT ✦

You are His poem, shaped with purpose, strengthened by grace, and formed with beauty in mind.

✦ REFLECTION QUESTIONS ✦

When have you felt most aligned with God while creating something? Be **specific.**

____

____

Where do you see design not just as utility, but as beauty?

____

____

How might you live differently today if you saw yourself as God's living poem?

____

____

✦ FAITH AND ACTION ✦

- Write a poem or a paragraph of prose about your relationship with God and with humans. How your Faith and Action have helped you in your rhythm of relationships.
- How you can live differently as God's living poem.
- Include what you've experienced through this day that's beautiful in life.
- A major element of good poetry and prose: Being Specific!

✦ NOTES TO GOD ✦

Week 3

ROCKET SCIENCE & ASTRONOMY

Motion, Momentum, and a Higher Perspective

We've explored how the universe is structured and how systems are designed. Now, we lift off into motion, velocity, and **perspective.**

Rocket science and astronomy push us beyond the surface. They ask how far we can go, how fast we can move, and what it takes to break free from what holds us down. This week is about **movement**: thrust, orbit, escape velocity, and the view from above. Yes…Action!

But this isn't just about propulsion.

It's about **perspective**. It's about **Faith**!

Because when astronauts look back at Earth from space, they don't just see data, they see **a story.** They feel awe, wonder, and reverence. They feel **small... and significant** all at once.

That's what the arts do, too.

Where science measures acceleration, **music measures movement and emotion.** Where math models velocity, **poetry captures stillness and spirit.** Where physics explains gravity, **painting displays a work of art, capturing what it feels like to be grounded, or what it feels like to rise.**

So yes, this week is about speed, force, and celestial bodies. But it's also about the God who gives YOU **eyes to see differently**, and a heart that can turn data into **devotion**.

Because **faith needs lift. And art gives it wings.**

Day 15

ESCAPE VELOCITY

Passage: Romans 6:14

[14]For sin shall no longer be your master, because you are not under the law, but under grace.

In rocket science, **escape velocity** is the minimum speed needed for an object to break free from the gravitational pull of a planet or other body.

For Earth, that velocity is about **11.2 kilometers per second** (or roughly **25,000 miles per hour**). Anything less and you'll fall back. But if you hit that threshold? You're free. You've left the pull behind.

Spiritually, we all live under gravity too. Not the kind that keeps your feet on the ground, but the kind that tries to pull you back into patterns, mindsets, and sins you've already been delivered from.

Habits you thought were gone. Doubts you thought were settled. Shame you thought was buried.

They pull. Hard.

But here's the good news: **grace gives you escape velocity.** Not willpower. Not performance. Not religious effort. **Grace.**

Romans 6:14 says sin is no longer your master, not because you finally figured it out, but because you are no longer under law. You are filled with grace.

That's the fuel.

Merriam-Webster Dictionary defines grace: "unmerited divine assistance given to humans for their regeneration or sanctification; a virtue coming from God".

Grace doesn't just help you resist gravity. It empowers you to break free from it. It gives you the strength to launch into a new orbit, beyond shame, beyond fear, beyond the lies that say, "you'll always come back to this."

You don't have to circle the same sin forever.

You don't have to fall back.

You were made to rise. And God's grace propels that rising.

✦ FAITH FORMULA ✦

Grace isn't just forgiveness; it's propulsion. It's how you break free and stay free.

✦ GOD'S CONSTANT ✦

He doesn't hold you down, He lifts you out. Every. Single. Time.

✦ REFLECTION QUESTIONS ✦

Where in your life do you feel the pull of something you've already been freed from? Be **specific.**

____

____

Are you trying to escape on your own fuel or relying on the grace of God?

____

____

What would living beyond the pull of shame look like this week?

____

____

✦ FAITH AND ACTION ✦

- Be still. Listen. Feel. Can you experience God's grace lifting you? If you don't, pray. Ask God to propel you. In some simple action to do His will. To be free of past sins. To be caring for yourself and for others. To be of love and service in simple ways.
- If you do feel God's grace lifting you, ask for guidance in doing His will. Let Him show you the way.
- At the end of the day, write in your Notes to God how you did today. Were you honest and kind to yourself? To others? In what ways, however small, were you helpful?
- Where were you tempted to fall back into old fearful, selfish ways. Did grace help you? If not, ask God to lift you from those bonds. If grace did help, write a note of thanks.

✦ NOTES TO GOD ✦

Day 16

THRUST AND DIRECTION

Passage: Proverbs 16:9

9In their hearts humans plan their course, but the Lord establishes their steps

Thrust is what makes a rocket move. It's the force that pushes it forward, created by engines that expel mass at high speed in the opposite direction. It's not just power, it's **purposeful power**. But thrust without **direction** is dangerous. It's not enough for a rocket to launch, it has to launch **with aim**.

The same is true in your life. You might have passion. You might have gifts, drive, vision, energy, and ambition. But if your force isn't focused, you'll waste fuel, and worse, you could end up spinning in circles, burning out without ever reaching the destination God designed for you.

Proverbs 16:9 reminds us of this holy balance:

We plan.

God establishes.

We bring the thrust.

He provides the trajectory.

This isn't about sitting back and doing nothing. It's about trusting that **God's guidance will meet your effort**. It's about submitting your momentum to the One who sees the whole flight path.

In mission control, direction is determined by precise math and constant mid-course corrections. In faith, direction is refined through prayer, surrender, Scripture, community, and listening and responding to the quiet nudges of the Holy Spirit.

The question isn't "Do I have enough push?" The question is: "Am I pushing in the right direction? In God's direction?"

✦ FAITH FORMULA ✦

Thrust without guidance wastes fuel. Direction without trust stalls the mission.

✦ GOD'S CONSTANT ✦

He doesn't just light the fire, He sets the course.

✦ REFLECTION QUESTIONS ✦

Where in your life are you exerting energy without clear direction? Be **specific.**

____

____

Have you invited God to guide your steps or are you running on momentum alone?

____

What would change if you gave God permission to redirect your thrust?

____

____

✦ FAITH AND ACTION ✦

- Start the day writing in your Notes to God: What's your plan of action for the day? Be specific.
- How can God direct you? Be specific.
- Pray that God will direct you. Then move forward with faith and action.
- Find time to pause briefly during the day and check in with God. Are you still on track? If the day's not going as planned, ask God to redirect you with faith and honest words and action.
- At the end of the day, write in your Notes to God. How did you do, you and God? How did you adjust to unplanned change? Did you respond with self-will and plowing ahead? Or did you let God guide you?
- What has occurred through the day and night that you can be grateful for?

✦ NOTES TO GOD ✦

Day 17

FUEL EFFICIENCY

Passage: Hebrews 12:1

1...let us throw off everything that hinders and the sin that so easily entangles. And let us run with perseverance the race marked out for us.

Rockets need fuel. Lots of it. But they can't carry unlimited reserves, so every ounce counts. That's why engineers are obsessed with fuel efficiency: using the least amount of energy to achieve the greatest possible result.

Every drop has to matter.

Every gram has to justify its weight.

Now think about your life. Your time. Your energy. Your emotional bandwidth. How much of what you're carrying is actually weighing you down rather than moving you forward?

Hebrews 12:1 urges us to throw off everything that hinders us. Not just sin, but anything that slows us down, distracts our minds, or dilutes our focus. In other words: travel light. Run lean. Fuel wisely.

But let's go deeper.

Because sin, especially internal sin, can quietly devour your energy and joy.

- Fear hijacks your system and makes you loop the same orbit over and over.
- Anger is expensive. It burns fuel with no return.
- Pride overloads your mass with a false image you were never meant to carry.
- Unforgiveness keeps you tethered to things you should have released lightyears ago.
- Bitterness corrodes your engine from the inside out.

We often burn out not because we're doing bad things, but because we're ignoring the drag of these sins, drag that slows us down, stifles our calling, and blocks the flow of grace and clarity.

Here's the beauty:

God is not just your source of power. He's your strategist.

He shows you not just how to move, but how to move cleaner. How to confess. How to heal. How to burn holy and run free.

Fuel wisely.

Burn clean.

And run the race that's actually yours to run.

✦ FAITH FORMULA ✦

Sin drains what grace empowers. Efficiency starts with internal freedom.

✦ GOD'S CONSTANT ✦

He doesn't just forgive sin, He frees you from its weight.

✦ REFLECTION QUESTIONS ✦

What sin or emotion is quietly burning fuel in your spirit? Be **specific.**

____

____

How can you confess it and release it into God's hands today?

____

____

What would it feel like to move through life lighter, cleaner, and freer?

____

____

✦ FAITH AND ACTION ✦

- Start today and every day with pausing to Get honest with God: Are you afraid? Or do you have faith? Or both?
- What are you afraid of? Be specific. If you're confused and don't know, talk to God about it. Ask for clarity, understanding, and direction.
- If you know what you're afraid of, ask God to lift that fear and fill you with His faith and action. With His insight and caring. With His loving sense of humor. It's all there for the asking.
- At the end of the day, write in your Notes to God. Did you move through life lighter, cleaner, freer? Why or why not? Get honest with God so He can help you. In. All. Things.
- If we do this, guess what? Reality comes. Clarity comes. Freedom from fear and confusion comes. And gratitude is not far behind.

✦ NOTES TO GOD ✦

Day 18

ORBITS AND DRIFT

Passage: John 15:4

4Remain in me, as I also remain in you. No branch can bear fruit by itself; it must remain in the vine....

In space, it's not enough to launch a rocket. At some point, that rocket has to settle into a **stable orbit**, a delicate balance between speed and gravity that keeps it circling the Earth (or any object) without crashing into it or flying away.

Too fast? You drift.

Too slow? You fall.

Too distracted? You miss your window.

Astronauts and satellites rely on this precision. **Orbit is not a one-time decision, it's a constant balance and flow.**

So is your walk with God.

We don't just "launch" into faith once and coast forever. The English Stand Version of the Bible begins

John 15:4 with "Abide in me,." so we are called to **abide**, to remain in Him, stay connected, stay in rhythm. Because just like in space, spiritual **drift is real**.

- Drift happens when you stop praying regularly.
- Drift happens when you stop reading Scripture and let other voices set your direction.
- Drift happens when you get hurt, pull away, and never circle back.
- Drift happens slowly. Quietly. Then suddenly.

  One day you wake up and realize you're far from the center, not because you made a big rebellious move, but because you stopped checking your alignment.

But here's the hope:

You can recalibrate.

Just like engineers fire small thrusters to re-center a drifting satellite, you can return to God with a simple prayer: *Lord, bring me back.*

Jesus didn't say, "Visit me when you feel like it." He said, *"Remain in me."*

That's how fruit grows.

That's how peace stays.

That's how drift becomes devotion and balanced progress again.

✦ FAITH FORMULA ✦

Abiding is active. Without connection, drift is inevitable. But God always welcomes you back.

✦ GOD'S CONSTANT ✦

He never moves. When you're ready to return, He's already there.

✦ REFLECTION QUESTIONS ✦

Are there signs of spiritual drift in your life right now? Please explain and be **specific.**

____

____

What practices help you stay centered in God's presence?

____

____

How can you realign today, not perfectly, but intentionally?

____

____

✦ FAITH AND ACTION ✦

- Start your day by **getting specific**: What **concrete** practices help you stay centered through the day and night…in God's presence? Write them in your Notes to God. Now ask Him to guide you through the day and night. To provide the grace for those concrete practices.
- At the end of the day, check back in your Notes to God: Did you and God perform those practices? If so, or if not, what was the result?
- Did you experience times you drifted away? If so, admit it. God already knows it. And God's ready to bring you back into balanced orbit. You need only ask. He knows, for us, it's progress, not perfection.

✦ NOTES TO GOD ✦

Day 19

GRAVITY WELLS

Passage: Psalm 40:2

2He lifted me out of the slimy pit, out of the mud and mire, he set my feet on a rock and gave me a firm place to stand.

In astrophysics, a **gravity well** is a region of space where the gravitational pull of an object like a planet, star, or black hole is so strong that it becomes extremely difficult to escape.

The deeper the well, the more force is required to break free. The closer you get to the center, the more energy it takes to turn around. And if you're too deep, you can feel like you're stuck for good.

Sound familiar?

We may not be orbiting stars, but we know what gravity wells feel like.

- The pull **of addiction**.
- The **weight of shame**.
- The **cycle of self-doubt**, or the **grip of depression**.

- **The repetition of sin** you swore you'd never return to.

You start trying to climb out, only to slide back in. The pit feels deeper every time.

But Psalm 40 reminds us: **God is not afraid of deep wells.**

He lifts.

He rescues.

He steadies your feet where there was once only sinking sand.

You might not have the escape velocity on your own, but **grace does**. And here's the truth: *the deeper the gravity well, the greater the glory when God pulls you out.*

No pit is too deep. No pattern is too strong. No mistake is too permanent. The same God who created gravity knows how to reverse it.

✦ FAITH FORMULA ✦

When sin pulls hard, grace pulls harder. What holds you down can't hold back God's rescue.

✦ GOD'S CONSTANT ✦

Even when you fall deep, He is still deeper and strong enough to lift.

✦ REFLECTION QUESTIONS ✦

What are the "gravity wells" (the things pulling you down) in your life right now? Please explain and be **specific.**

____

____

Have you been trying to escape by your own strength?

____

____

What would it look like to surrender to God's rescue today?

____

____

✦ FAITH AND ACTION ✦

- As you start your day, sit quietly with God for a few minutes. Write to Him in your *Notes to God.* Are you feeling the weight of a gravity well? Is fear or shame or self-doubt holding you back? Making you reluctant to take some positive action at home or at work. Making you feel stuck? Be specific and ask for God's help and grace.
- Maybe you're feeling wonderful this morning. Feeling the results of faith and honest action and trust in God. If so, tell Him. Share your gratitude with God.
- During the day and night, if you experience the pain of a "gravity well", you know where to take it, don't you? Let God fill you with grace and lift you. It's there for the asking.
- At the end of the day, report back to God how His grace has helped you.

✦ NOTES TO GOD ✦

Day 20

T-MINUS AND TRUST

Passage: Ecclesiastes 3:1

[1] There is a time for everything, and a season for every activity under the heavens.

If you've ever watched a rocket launch, you know the tension of the **countdown.** Everything is scheduled down to the second. **T-minus 10. 9. 8.** And then suddenly…

Hold.

Mission control pauses the launch. Maybe there's a weather issue. Maybe it's a minor system error. Maybe it's something invisible to the public eye but visible to the engineers who know what's at stake.

And here's the thing: the astronauts in the capsule don't always know why the launch is delayed.

They've trained.

They've prepared.

They're ready to go.

But they still have to **wait** on the voice from mission control.

Sound familiar?

Maybe you've been preparing, praying, pressing, *ready to launch* and God says, *"Hold."*

- That job opportunity you were sure would open up.
- That relationship you were sure was "the one."
- That door you've knocked on a hundred times.
- That dream that felt so close you could touch it.

But instead of liftoff, you're sitting on the pad… engines ready… but waiting.

And perhaps the word comes: Not today. So, you exit the capsule.

But you don't give up on the mission. Your faith and work and connection to mission control lets you know: *We keep moving forward.*

Ecclesiastes 3 reminds us that **God's timing is not our timing.** And that's not a delay. That's design.

Because the same God who called you is the God who sees what you can't see.

He knows when you're really ready.

He knows when the atmosphere is just right.

He knows how to avoid failure by aligning you with the fullness of His perfect timing.

Trust the countdown.

Trust the pause.

God doesn't cancel launches. He perfects them.

✦ FAITH FORMULA ✦

A delayed launch isn't a denied mission. Trust the silence between countdowns.

✦ GOD'S CONSTANT ✦

When your clock says "wait," God is still working behind the scenes.

✦ REFLECTION QUESTIONS ✦

Where in your life do you feel stuck on the launch pad? Be **specific.**

____

____

How are you responding to God's "hold" moments?

____

____

What does trust look like when you don't have a launch date?

____

____

✦ FAITH AND ACTION ✦

- Start your day by talking to God about what you see as your mission **today**.
- If you're not sure, ask for His help and guidance. If you are sure, *be specific* in asking for His help and guidance. He is mission control. We may be the pilots, but He designed the flight plan. He sees the entire course as it truly is.
- As you move through your day and night, trust mission control. Trust the pauses. Trust the reality of perhaps having to change the launch, the flight course. Realize that God sees it all and wants to guide you.
- At the end of the day, write in your Notes to God how your day and night went. Did you let Him guide you? Where were you tempted to fearfully, willfully take control? Did you try, or did you trust God? Let Him know how it went. If it went well, consider thanking Him. If it didn't, consider asking Him to guide you, so tomorrow you can conduct the mission according to His plan.

✦ NOTES TO GOD ✦

Day 21

THE OVERVIEW EFFECT

Passage: Isaiah 55:9

9 As the heavens are higher than the earth, so are my ways higher than your ways and my thoughts than your thoughts.

When astronauts leave Earth and look back at it from space, something powerful happens. It's called **the overview effect**, a cognitive shift in awareness caused by seeing the planet from orbit or the moon.

They see Earth not as borders, battles, or busy schedules, but as a fragile blue marble, suspended in darkness, unified and vulnerable. From that distance, things that once felt massive suddenly look small. The big picture becomes clearer than ever before.

Perspective changes everything.

The same is true in our walk with God. We live most of life zoomed in, handling bills, fielding texts, chasing goals, managing relationships, and navigating emotions.

From that up-close view, small things feel overwhelming. We get anxious, reactive, or discouraged.

But what if we let God lift our gaze?

Isaiah 55:9 reminds us that God's thoughts are higher. Not just bigger, but *truer*. When we step back and let God reframe what we're facing, we begin to see with **kingdom eyes**. That doesn't mean we ignore the details, but it means we view them through a lens of hope, eternity, and divine sovereignty.

The overview effect isn't denial. It's clarity.

And you don't have to leave the planet to get it.

Sometimes all it takes is a moment of stillness, a prayerful breath, or time in the Word where God shows you: *I'm still in control. I still see you. And your story is part of something far greater than this moment.*

✦ FAITH FORMULA ✦

Perspective is power. When you see from God's view, everything else aligns.

✦ GOD'S CONSTANT ✦

He's never stuck in the weeds. He holds the whole picture, and He's still holding you.

✦ REFLECTION QUESTIONS ✦

What situation in your life needs a perspective shift right now? Be **specific.**

____

____

How can you create space today to "zoom out" and ask God for clarity?

____

____

What changes when you view your struggle as part of a bigger story?

____

____

✦ FAITH AND ACTION ✦

- Start today by writing to God: What's your perspective of yourself? Of your relations to loved ones? To fellow workers? Be specific.
- Now ask God to grant you His perspective: bring His *reality* into your mind and heart so you can see what's true. Not what fear or expectation might lead you to see.
- Ask God to guide you: to help you move through the day and night with *honest words and actions.* With kindness and love for yourself and others.
- Be willing to forgive yourself and others for being human. We're not God. We're not perfect. This perspective, this reality helps infuse love into all our relationships.
- At the end of the day, write in your Notes to God: Did your perspective shift as you moved through your day and night? Were you able to speak and act honestly, lovingly? Be specific. Let God know so you both can see your true relationship. How you're growing and changing.
- Write a note of thanks!

✦ NOTES TO GOD ✦

CREATIVE SPARK
(Week 3 Close)

THE ART OF SEEING FROM ABOVE

(as seen through the lens of the creative spark)

Passage: Isaiah 40:26 (ESV)

26 Lift up your eyes on high and see: who created these? He who brings out their host by number, calling them all by name...

There's a moment many astronauts describe after leaving Earth: a deep shift in consciousness known as the **Overview Effect**. From space, politics fade. Borders disappear. And what they see is not power, but **poetry**.

A blue marble spinning in silence.

A swirl of cloud and color.

A place called home.

Science may call it a shift in perspective.

But artists have always called it by what it is: **vision**.

We don't have to leave the atmosphere to experience it. Poets, painters, and photographers access this kind of clarity all the time, not by changing location, but by **changing how they see.**

Perspective is *spiritual.*

It's the difference between seeing stars as data points, or as declarations of God's glory.

It's what allows a photographer to see a whole story in a shadow.

What leads a songwriter to capture the ache of waiting in just four chords.

What lets a dancer translate grief into grace.

This is the **art of looking from above.**

Not just observing the moment but transcending it.

Not just analyzing the problem but imagining something new.

Isaiah tells us to *"lift our eyes on high and see..."* because God isn't just calling you to move through the world. He's calling you to **see it differently.** To rise. To reframe. To receive beauty as part of the truth.

Because sometimes transformation doesn't begin with action.

Sometimes it begins with a holy pause... and a better view.

✦ FAITH FORMULA ✦

Perspective is prophetic. Art invites us to see the world not only as it is, but as it *could* be.

✦ GOD'S CONSTANT ✦

The Creator of the stars is also the Artist of your perspective, and He's always inviting you to look again.

✦ REFLECTION QUESTIONS ✦

Where are you being called to step back and see the bigger picture? Be **specific.**

____

____

What form of art helps you shift your perspective when you feel stuck?

____

____

How might seeing the world "from above" change how you engage it below?

____

____

✦ FAITH AND ACTION ✦

- Sit, relax, and try some *stream-of-*conscious writing: Just write down whatever flows through your mind.
- Now, write down *specific images* to describe your perspective: Of God. Of the world. Of yourself. Of loved ones. Of fellow workers.
- Look at what you've written. What does it tell you about your perspective on your life? Where do you see room for growth? For change? For gratitude?
- Consider following the same process with art on other nights: *Draw or paint* images of your perspective. See where it takes you.
- Celebrate the creative process as God celebrates it!

✦ NOTES TO GOD ✦

Week 4

MATH, DATA, AND DIGITAL LIFE

Patterns, Precision, and the Presence of God in the Code

If rocket science teaches us how to move, **math** teaches us **how to measure**. If engineering shows us how to build, **data** shows us **how it works.** And in the digital world that we now live in, algorithms, code, and artificial intelligence shape the rhythm of daily life more than most of us even realize.

But math and data are not code, they're **designed.**

They reflect a God of order, pattern, logic, and **mystery**.

This week, we'll explore:

- **Pi**, the number that never ends
- Algorithms, and what it means to follow the instructions of God
- Limits, and the beauty of a limitless Creator
- Error **correction**, and how grace recalibrates us

- Binary **code**, and the image of God hidden in the simplest digital language
- Exponential **growth**, and how small things multiply in the Kingdom
- And finally, **AI**, and what it means to follow "I Am" in a world of artificial minds

Whether you're a math lover or someone who breaks out in a sweat at the mention of long division, these next days will show you that **God speaks through numbers, logic, and code**, and that His voice still cuts through the digital noise.

And in the midst of all this structure, you'll find the **creative spark.**

Because music is math.

Poetry has patterns.

Architecture obeys ratios.

And great storytelling follows sacred geometry.

This week don't be afraid of structure.

Step into it. Then let God show you where the numbers give way to **nuance.**

Where formulas give way to **faith.**

And where code becomes **communion.**

Day 22

THE BEAUTY OF PI

Passage: Romans 11:33

33 Oh, the depth of the riches of the wisdom and knowledge of God! How unsearchable his judgments, and his paths beyond tracing out!

Pi (π) is one of the most famous numbers in the universe. It begins with **3.14159...** and keeps going, infinitely. It never ends. It never repeats. It simply **is**. And yet, it shows up in some of the most ordinary and extraordinary places such as circles, waves, planets, pendulums, signal processing, architecture. Anything that moves in rhythm or curves with elegance… it carries a trace of Pi.

It's almost like creation has God's fingerprint baked into its geometry.

And the most amazing part?

Even after billions of digits have been calculated, we still haven't found a pattern. **Pi is endless, but also full of mystery.**

Just like God.

Romans 11:33 captures this beautifully: *"How unsearchable His judgments, and His paths beyond tracing out."* You can calculate, question, explore, and go deeper, and you should. However, you'll never get to the end of God.

He is not a math problem to be solved.

He's a mystery to be worshiped.

And like Pi, He's always present in the pattern, even when you can't quite trace it.

This truth does something powerful: It frees us from needing to "figure it all out" before we follow Him. It reminds us that awe is just as holy as understanding. And it affirms that we can trust a God who is not small enough to be fully explained, but **near enough to be experienced**.

✦ FAITH FORMULA ✦

God isn't just infinite, He's intentional. The mystery is part of the majesty.

✦ GOD'S CONSTANT ✦

He never ends. He never repeats. But He is always the same.

✦ REFLECTION QUESTIONS ✦

Where are you trying to fully understand something God might be asking you to simply trust? Be **specific.**

____

____

How do you respond when God feels mysterious rather than clear?

____

____

What would it look like to worship God for His depth, not just His answers?

____

____

✦ FAITH AND ACTION ✦

TODAY, BEGIN WITH a simple prayer: "God, please let me trust You today."

As you walk through your day and night, feel free to repeat this prayer. See where it takes you. See how you respond as you plan your day.

Let God's infinite love and wisdom guide you through the problems you feel you can't solve. As you experience success or disappointment. Or if a negative co-worker confronts you. Let your action be this: Trust God, then respond. Can you see God in yourself and the other person? Or is fear, your refusal to trust, holding you back?

At the end of the day, write in your Notes to God. Where did you trust Him through your day and night? Where were you afraid to? What was the result?

Consider closing your writing with a note of gratitude for His gift of trust. And a prayer that it will grow deeper as you walk with Him.

✦ NOTES TO GOD ✦

Day 23

ALGORITHMS AND OBEDIENCE

Passage: Proverbs 3:5-6

5 Trust in the Lord with all your heart and lean not on your own understanding; 6 in all your ways submit to him, and he will make your paths straight.

An **algorithm** is a set of instructions. Step-by-step processes that tell a system exactly what to do to solve a problem, make a decision, or reach a goal.

Algorithms run everything from search engines to self-driving cars. They don't ask why. They just follow the code.

Now, that doesn't mean God wants robotic obedience. He gave us free will, curiosity, and creativity. But there's something powerful, **even freeing**, about the idea of trusting God's instructions like an algorithm.

You don't have to understand the entire plan.

You don't have to write the code yourself.

You just have to **follow the next step**.

When God gives a prompt, be it **forgiveness, generosity, rest, boldness, surrender,** He's not asking for your full analysis. He's asking for your trust. Proverbs 3 says it plainly: *"Lean not on your own understanding."*

Because the truth is: our understanding is limited.

We don't see the whole dataset.

We don't know the full future.

But God does.

He's not just writing your story. He's optimizing it.

He's not just programming outcomes. He's shaping your heart with each step. What do you do when you don't know what to do next? Obey the last thing He told you. Trust the code. **And follow the Designer.**

✦ FAITH FORMULA ✦

Faith isn't knowing every line of code. It's obeying the next instruction.

✦ GOD'S CONSTANT ✦

His directions are never random. Each step is part of something bigger.

✦ REFLECTION QUESTIONS ✦

What's the "next instruction" you sense God giving you right now? Be **specific.**

____

____

Are you hesitating because you don't understand the full outcome?

____

____

What would it look like to follow in obedience, trusting that God wrote the full script?

____

____

✦ FAITH AND ACTION ✦

TODAY, ALIGN YOURSELF with God's algorithm, his basic set of instructions: Faith, Hope, and Charity.

Now, the big test for us: Do we love all humans for the sake of God? We can love them ALL because God loves them ALL. If we doubt we do, or feel we can't, even more reason to ask for His grace, his infused virtue of love. It's the key to our living as one with God.

At the end of the day, write in your Notes to God: Did you activate them in your care for yourself and others? In your service to others? This is not about perfection. It's about moving forward in loving the best we can, knowing God loves us.

✦ NOTES TO GOD ✦

Day 24

THE LIMIT DOES NOT EXIST

Passage: Ephesians 3:20

[20]Now to him who is able to do immeasurably more than all we ask or imagine, according to his power that is at work within us.

In calculus, a **limit** defines how a function behaves as it approaches a certain point. It helps map expectations, contain behavior, and create structure.

But then there's the **asymptote**, a boundary line that a curve approaches but never quite touches. It gets closer and closer, infinitely near, but never intersects.

An asymptote says: *You can't reach this, but it shapes everything you move toward.*

Now, imagine your understanding of God.

Ephesians 3:20 emphasizes, He can do *immeasurably more* than all we ask or imagine. That means **God is your divine asymptote**, always near, always shaping, always drawing you deeper into mystery, glory, and awe. And yet, no matter how far you grow, there's still more.

He is:

- Beyond the limits of your intellect
- Beyond the borders of your expectation
- Beyond even your most faith-filled prayers

The more you know Him, the more you realize you've only just begun. The more you trust Him, the more He reveals new depths of His goodness. The closer you get, the more clearly you see, **there is no end to Him.**

God is not bound by the edges of your comfort zone. He doesn't work within your box. He is infinite, and He invites you to grow infinitely toward Him.

✦ FAITH FORMULA ✦

God is the divine asymptote, always drawing you closer, always giving more, and never running out.

✦ GOD'S CONSTANT ✦

You can never exhaust His presence, His love, or His power. He is immeasurable and always near.

✦ REFLECTION QUESTIONS ✦

What are you approaching right now that feels "just out of reach"? Be **specific.**

____

____

Could it be that God is using that very pursuit to draw you closer to Him? Please explain.

____

____

Where is He inviting you to embrace wonder rather than demand a finish line?

____

____

✦ FAITH AND ACTION ✦

EARLY IN YOUR DAY, perhaps when just leaving for work, walk outside and stand still for a moment. Look up at the sky. Look closely at how infinite it is. Clouds and blue and a distant sun. And more beyond what we can see. And if it's a really cloudy day...certainly more there than we can see.

Now, consider a prayer: "God, you are all this and beyond this. And I am here within all this and within You. Please guide me through your day and night. Help me pay attention. Show me the way."

At the end of the day, write in your Notes to God how your day and night went. What did you do? Push forward? Or pause, and ask, "Show me the way." What results did you experience?

✦ NOTES TO GOD ✦

Day 25

FAIL FORWARD

Passage: 2 Corinthians 12:9

9But he said to me, "My grace is sufficient for you, for my power is made perfect in weakness."...

In Science and engineering, failure isn't just expected, it's *built in.* Prototypes flop. Equations break down. Systems crash in testing. And yet, each failure offers something valuable: **information.**

Engineers don't fear failure. They **study it**. They learn from it, refine the process, tweak the inputs, strengthen the design. Because what doesn't work is often the fastest path to discovering what *does.*

In the Kingdom of God, it's no different. You are going to mess up. Miss deadlines. Say the wrong thing. Fall into old habits. Disappoint people. Drop the ball.

But here's the thing: **failure doesn't disqualify you.**

Not in God's design.

God doesn't expect perfection. He expects dependence.

Paul learned this the hard way. He wanted a weakness removed, but God said, *"My grace is sufficient. My power is made perfect in weakness."* That means your failure may be the very place God is most present, **not in spite of it, but through it.**

He's not tallying your breakdowns, He's transforming them. What you see as "the end of your capacity," God sees as **the beginning of His power.**

Stop disqualifying yourself for being human. Instead, bring your failure to the table, and let it become **fuel for formation**.

Fail.

Learn.

Adjust.

Repeat.

That's how engineers build rockets. And it's how God builds saints.

✦ FAITH FORMULA ✦

Failure isn't the end of faith. It is often where real faith begins.

✦ GOD'S CONSTANT ✦

He doesn't discard you when you fall, He develops you when you return.

✦ REFLECTION QUESTIONS ✦

What recent failure has caused you shame or self-doubt? Please explain and be **specific**.

____

____

How might God be using that very moment to refine your trust or clarify your purpose?

____

____

What would it look like to bring your weakness boldly to Him today?

____

____

✦ FAITH AND ACTION ✦

AS YOU MOVE through your day and night, in that moment when something's gone wrong, take the *action of pausing*. Pray.

Ask God to show you the reality: "What's really happening here, God? This is not what I expected. Why is this not working? What should WE do now, You and I? Guide me in actions to set this on Your path. Help me to accept my weakness and Your strength, Your perfection. To walk with You, that I may know and do Your will."

At the end of the day, write in your Notes to God: What went wrong? Did you pause? Did you ask for His guidance? If not, why not? If you did, how did it turn out?

✦ NOTES TO GOD ✦

Day 26

BINARY AND BELOVED

Passage: John 1:1

[1]In the beginning was the Word, and the Word was with God, and the Word was God.

Binary code is the foundation of everything digital. At its core, it's a language made up of only two digits: **ones and zeroes**. But here's the beauty, when those ones and zeroes are **combined in different patterns**, they can represent letters, numbers, images, sounds, and entire digital experiences.

For example:

- The letter "A" is 01000001 in binary.
- The number "5" is 00110101.
- Your favorite song or the image of your face on a screen? All made possible by different combinations of those two simple digits.

It's astonishing: **from binary comes incredible complexity.** What seems basic on the surface powers the most advanced systems in the world.

Now consider this:

In the digital world, your identity is often reduced to binary, data points and digital footprints.

But in the Kingdom of God?

You are **not a string of code. You are Beloved.**

You're not defined by analytics, clicks, or algorithms.

You are known by name, by heart, by design.

John 1:1 says, *"In the beginning was the Word."*

That Word, the **divine code** of life itself, wasn't a program. It was a Person.

Jesus.

And the same God who spoke galaxies into being also **formed your soul** with intention and image-bearing dignity. You're not a machine. Not a mistake. Not a metric.

You are His.

We may live in a world of screens, systems, and signals, but there is a deeper reality: **God didn't generate you, He handcrafted you.** He doesn't process your worth, He **declares it.**

✦ FAITH FORMULA ✦

You may operate in a binary world, but you are wired by love, not logic.

✦ GOD'S CONSTANT ✦

You are not a number. You are a name. You are seen, known, and chosen.

✦ REFLECTION QUESTIONS ✦

Have you ever felt reduced to your performance, output, or data?

____

____

What does it mean to you that God sees you as fully human and deeply beloved?

____

____

How can you unplug from comparison and reconnect to your God-given identity?

____

____

✦ FAITH AND ACTION ✦

KEEP IT SIMPLE TODAY. Start with this prayer: "Good morning, God. It's me, (SAY your name.) Thank You for loving me. Just as You love me, guide me in loving others throughout. Please let love and service be at the center of all I do today."

At the end of the day, write in your Notes to God how your day and night went. Be specific. Where were you and God able to be of love and service. Where did you fall short. Know that falling short is not failure. It's being human, and God opening an avenue for growth and loving Him and others even more. Know that He loves you for your faith and action.

✦ NOTES TO GOD ✦

Day 27

EXPONENTIAL GROWTH

Passage: Matthew 13:31-32

[31] *... The kingdom of heaven is like a mustard seed, which a man took and planted in his field.* [32] *Though it is the smallest of all seeds, yet when it grows, it is the largest of garden plants and becomes a tree..."*

In math, **exponential growth** occurs when something multiplies at an increasing rate. It starts small, barely noticeable, but it doubles, triples, and expands until the curve shoots upward like a rocket.

You've seen it in action:

- One spark starts a wildfire.
- One cell becomes a living organism.
- One video goes viral and reaches millions in hours.

Exponential growth says: **small beginnings are not small endings.**

Jesus understood this. That's why He used the mustard seed to describe the Kingdom. *"Though it is the smallest of all seeds...*" It's easy to overlook. Easy to dismiss. But in God's hands, the small becomes **significant**.

That idea applies to your faith, too.

- That one prayer you pray in secret? It echoes across generations.
- That one act of kindness? It ripples through someone's entire week.
- That one seed of obedience you plant today? It might bloom into something bigger than you'll ever see in your lifetime.

We live in a culture obsessed with **instant results and viral growth**. But Kingdom math works differently. It honors the slow build, the unseen root system, the early mornings and late nights, the faithful investment when no one's clapping.

And when the time is right, **the curve turns.**

Growth explodes. Impact multiplies. God does what only He can do: He takes your little and turns it into **legacy.**

✦ FAITH FORMULA ✦

Plant in faith, water with obedience, and trust God with the multiplication.

✦ GOD'S CONSTANT ✦

He sees the seed long before the tree. And He never forgets what you've sown.

✦ REFLECTION QUESTIONS ✦

Where in your life are you underestimating the power of a small beginning? Be **specific.**

____

____

What "mustard seed" have you been called to plant today?

____

____

Are you willing to trust God for the increase, even if you never see the full harvest? Please explain.

____

____

✦ FAITH AND ACTION ✦

START YOUR DAY with your Notes to God. Write these simple words:

"Mustard Seed = Life"

"Mustard Seed = Me"

Now whisper this simple prayer: "God, please ***grow*** me today."

Carry that image of the mustard seed with you. With every person you see, say to *yourself,*

At the end of the day, write in your Notes to God how it went. Be specific.

✦ NOTES TO GOD ✦

Day 28

AI AND THE I AM

Passage: Exodus 3:14

[14]God said to Moses, "I AM WHO I AM. This is what you are to say to the Israelites: I AM has sent me to you."

Artificial Intelligence (AI) is everywhere. From virtual assistants that answer your questions to algorithms that write essays, trade stocks, generate artwork, and even mimic human conversation. AI can predict, respond, learn, and adapt. It's advancing fast, faster than most of us are comfortable with.

And yet, for all its power, all its speed, all its simulated intelligence… AI is still just that: **artificial.**

It operates on **input and code.**

It reflects its programming.

It learns only what it's allowed to learn.

And it has no soul.

Now, contrast that with the name God gives Himself in Exodus 3: *"I AM."*

No beginning. No end. No creator. No code. No limit.

Where AI says, *"I am what you programmed me to be,"* God says, *"I Am who I Am."*

AI can mimic intelligence, but it cannot possess wisdom. It can replicate voice, but it cannot breathe Spirit. It can scan the Scriptures, but it cannot hear God. **It can predict your behavior, but only God can transform your heart.**

As believers, we should engage technology with discernment and curiosity, but **never confusion**. The most complex AI model will never replace the authority, intimacy, or presence of the Great I AM.

So let AI help you schedule your day, but don't let it shape your soul. Let it generate ideas, but don't let it define your identity. Let it inform your decisions, but let **God** direct your path. Because when all systems fail, and the servers go dark, and the signal cuts out, **He still *is*.**

✦ FAITH FORMULA ✦

Artificial intelligence may process information, but only the I AM speaks life.

✦ GOD'S CONSTANT ✦

He was before data, beyond algorithms, and will outlast every system. Always present. Always personal.

✦ REFLECTION QUESTIONS ✦

Where are you tempted to let technology shape your pace, identity, or values? Be **specific.**

____

____

How does God's name, *I AM*, anchor you in a world that's constantly updating?

____

____

How can you keep your connection to the Source stronger than your connection to the system?

____

____

✦ FAITH AND ACTION ✦

IF YOU'RE READING THIS in a physical book, put it down. If you're reading this on a devise, put it down.

Sit quietly. Deep breathe to relax. Listen. Consider how whatever you hear…the house creaking…leaves rustling in the breeze…a car passing…voices speaking in the other room…the computer beeping…it is all God's creation. It and you are all part of His infinite continuum.

God's Spirit is in all, including you. Including every person you'll encounter. "I AM."

Carry that image with you throughout your day and night. See if it helps you when communicating with others. When something doesn't go as you planned. When the computer and AI give you limited, programmed information, know that God has infinite, clear information waiting for you if you pray and listen.

At the end of the day, write in your Notes to God if this worked for you. Did you let artificial intelligence guide you, or open to God's infinite intelligence and guidance? Be specific. Feel free to ask God to clarify whatever puzzles you. To show you the way in all things, since He creates all things. Since He IS ALL THINGS. "I AM."

✦ NOTES TO GOD ✦

CREATIVE SPARK

(Week 4 Close)

THE POETRY OF PRECISION

(as seen through the lens of the creative spark)

Passage: Ecclesiastes 3:11

[14]He has made everything beautiful in its time. He has also set eternity in the human heart..."

A well-written equation has the same impact as a well-written poem. It moves something in us. Not because we understand every variable, but because we recognize the elegance in the pattern. Math is not just numbers—it's the architecture of creation. And in that architecture, there is beauty.

There's beauty in a musical chord progression that follows a Fibonacci sequence. Beauty in the way Bach composed fugues using symmetry and math. Beauty in the repetition of poetry that mirrors the rhythm of breath. Beauty in the dance between precision and play.

We don't always think of math as spiritual. But it is.

- Every ratio in nature.
- Every orbit in the sky.
- Every fractal in a snowflake.
- Every beat in a line of spoken word.

They are all evidence that God didn't just create the world to function; He created it to flow.

Ecclesiastes tells us that God has set eternity in our hearts. That means that we long for what's infinite, but we also crave what's ordered. We are drawn to both freedom and form. That's why we're moved by both a spontaneous jazz solo and a perfectly timed haiku.

Creativity and structure are not opposites. They're partners. And when we honor the sacred geometry of a moment—whether through code, design, rhythm, or ratio—we echo the wisdom of a God who composed the cosmos with both calculation and compassion.

✦ FAITH FORMULA ✦

Structure doesn't stifle creativity—it reveals it. The Divine Artist moves through form to unlock beauty.

✦ GOD'S CONSTANT ✦

He designed you with rhythm, timing, and intentionality. Your life is part of His eternal composition.

✦ REFLECTION QUESTIONS ✦

Where do you see beauty in structure? Do you see it in your daily routines, in your design, and in your spiritual practices? Be **specific.**

____

____

How does knowing that God values both order and imagination shift how you see your work?

____

____

What is one area of your life where you can bring more intentional rhythm and creative energy?

____

____

✦ FAITH AND ACTION ✦

IN THE MORNING, write in your Notes to God section this title to a poem: **I See God and Beauty at Home**

Now sit quietly and pay attention to what you see, hear, and feel. Also, pay attention to what you imagine – the creative possibilities.

Now write a poem. What concrete images in your life at home matter to you? Where do you see God and beauty in them?

Then, at the end of the day, write this title to a poem: **I See God and Beauty Throughout My Life**

What images come to mind? Write them down. Consider including expressions of gratitude for your experiences of God and beauty. Don't worry about being perfect. Just ask God to guide you in experiencing Him and His beauty. See where it takes you.

✦ NOTES TO GOD ✦

Week 5

SYNTHESIS… *AND LAUNCH*

Integration, Urgency, and Stepping Boldly into Your Calling

We've studied the laws. We've explored the systems. We've lifted into orbit. We've found wonder in the numbers. Now it's time to bring it all together and **go**.

Week 5 is about **synthesis**, the coming together of all the parts. Faith, intellect, passion, and purpose aligned for movement. You'll read about system integration, holy urgency, and the call to live mission ready.

But this isn't just about utility. It's about wholeness. Because you weren't made to live compartmentalized. Your faith isn't one tab open in a crowded browser. It's the **operating system** that holds it all.

God doesn't just want your mind; He wants your imagination. He doesn't just want your obedience; He wants your offering. He doesn't just want you to do the work; He wants you to see the **art** in it.

This final week also includes one last reflection, as seen through the lens of the creative spark, which serves as a reminder that **launch** is not only about doing; It's also about **creating**. You were formed by the hands of the Creator. Now, you are sent to **create forward.**

Day 29

SYSTEM INTEGRATION

Passage: Colossians 2:6, 9-10

6So then, just as you received Christ Jesus as Lord, continue to live your lives in him...
For in Christ all the fullness of the Deity lives in bodily form, and in Christ you have been brought to fullness.

In engineering, **system integration** is the process of combining various subsystems, each with different functions, into one seamless, functional whole.

Imagine a rocket. There's propulsion. Navigation. Communication. Life support. Fuel systems. Sensors. All of them are essential. But without integration, they don't work together. The rocket might launch, but it won't survive the mission.

It's not just about having strong parts; It's about having one **coherent system.** Your life works the same way. You have gifts, goals, relationships, dreams, routines, responsibilities, and maybe even a growing faith. But without integration, without **anchoring everything in Christ,** it's easy to live in fragmentation.

You go to church, but you compartmentalize your business. You pray, but you don't surrender your ambition. You speak love, but you carry bitterness in secret. You function, but you're not **whole**.

Colossians 2 reminds us: "In Christ you have been brought to fullness."

Not fragments. **Fullness.**

Not separate systems. **One integrated life.**

Jesus isn't supposed to be one part of your system; He's the **source code**. The operating system. The integrator that holds every function of your life together and makes it flourish.

System integration isn't about perfection. It's about alignment. It's about living as one person, under one Lord, on one mission, without switching hats depending on who's watching.

Ask yourself today:

Are you functioning?

Or are you integrated?

✦ FAITH FORMULA ✦

Faith isn't one file in your system. It's the framework that makes the whole system work.

✦ GOD'S CONSTANT ✦

He doesn't just want part of your life. He came to bring wholeness to all of it.

✦ REFLECTION QUESTIONS ✦

Where are you living a fragmented life, divided between who you are and who you're called to be?

____

____

What would it look like to truly integrate your faith into your work, relationships, or decisions?

____

____

How is Jesus inviting you into fullness, not just function?

✦ FAITH AND ACTION ✦

THIS MORNING, take a moment to sit still. Ask God to guide you. Pray to be guided, and to **activate** the gift of Faith, Hope, and Charity as you move through your day of *love* and *service* at home, at work, and everywhere you go.

Pray to treat yourself and each individual as a Child of God…because each individual *is* a Child of God.

At the end of the day, write in your Notes to God on how you did. When was it difficult or impossible to experience? Ask God to guide you in growing within His Vision, and in activating it at all times.

✦ NOTES TO GOD ✦

Day 30

THE FINAL COUNTDOWN

Passage: Romans 13:11

11 And do this, understanding the present time: The hour has already come for you to wake up from your slumber, because our salvation is nearer now than when we first believed.

Every rocket launch ends with a countdown. You would hear, **"T-minus 10... 9... 8..."** Every second matters. Every team is focused. Every system is locked in. Because once that window opens, **you go.** There's no time to renegotiate the mission. No more testing. No more procrastinating. The final countdown is about **urgency**.

Paul understood this urgency when he wrote to the church in Rome: "Wake up… the hour has already come." Translation: **You're in the countdown. Stop waiting to start living.**

So many of us live as if we've got infinite prep time:

- "I'll get serious about my calling later."
- "I'll deal with my anger eventually."

- "I'll forgive when it feels easier."
- "I'll start the mission after I feel more qualified."

But the truth? **The window is now.** God is not waiting for you to be perfect. He's waiting for you to be present. There's a mission on your life that doesn't need to be imagined anymore. It needs to be **activated**.

You don't need more inspiration. You need movement. You don't need a bigger sign. You need obedience. Because when God says "go," **everything you need is already in motion.**

So don't sleep through your launch window. You were made to move. You were built to fly. You were wired for this exact moment in time.

✦ FAITH FORMULA ✦

You don't need more time. You need more trust. The countdown is confirmation that launch is near.

✦ GOD'S CONSTANT ✦

He doesn't rush, but He also doesn't waste time. His timing is precise, and you are right on schedule.

✦ REFLECTION QUESTIONS ✦

Where are you delaying action because you're waiting to feel "ready"? Be **specific.**

____

____

What's one step of obedience you know God is calling you to take today?

____

How would your choices shift if you lived like the countdown was real?

____

____

✦ FAITH AND ACTION ✦

THIS IS WHAT OUR EARLIER exercises have prepared us for: Faith in God. Trust in God. Love of God. Now let's move! Pray to God to guide us in moving forward. To carry these gifts to others in our thoughts, our feelings, our words, our actions. Pray to God to use ALL of ourselves in being of Love and Service EVERYWHERE to EVERYONE.

If that sounds like too gigantic a task, know this: You are not doing this alone. You are **never** alone. And you will not do it perfectly. God doesn't expect you to do it perfectly. He wants you to *propel* yourself with His Faith, Hope, and Love. To share it with the world and, yes, beyond the world. Together with God, you can do it.

At the end of the day, write in your Notes to God how you did. What worked well? What didn't work so well? Write it all down: your honest Notes to God. Your love letter to your Creator. Thank God for your shared life with Him.

✦ NOTES TO GOD ✦

MISSION READY

Passage: Matthew 28-19:20

[19]Therefore go and make disciples of all nations... [20]And surely I am with you always, to the very end of the age.

The rocket is built. The systems are aligned. The mission has been briefed. Now there's one last question: Are you ready?

In aerospace, "mission ready" doesn't mean you've figured everything out. It means that everything required to begin the journey is in place. It means you've completed your checks, accepted the risks, trusted the process, and locked into the call.

Jesus gave us our mission in Matthew 28: "Go."

Not "wait until it's perfect."

Not "collect more information."

Just "Go." Make disciples. Carry My name. Walk in My authority. That's your mission.

And through 31 days of reflection, science, scripture, surrender, and growth, you've been aligning with that call. Maybe you don't feel qualified. Maybe you still feel messy or unsure. But if God calls you, you're not only capable, you're commissioned.

Remember, astronauts don't launch themselves. They ride the thrust of something greater than them. You don't go in your own strength—you go in the power of the I AM who sends you.

Mission readiness doesn't mean no fear. It means more faith than fear. It means waking up with holy confidence that the same God who designed black holes and blood vessels, quantum fields and galaxies, is the same God who designed you with a purpose that matters.

So now it's your turn. Liftoff isn't a someday thing. It's today. You were born for this. You've trained for this.

You are mission ready.

✦ FAITH FORMULA ✦

You don't have to be flawless to be faithful. Mission ready is about surrender, not perfection.

✦ GOD'S CONSTANT ✦

He doesn't just send you. He goes with you. Always.

Even to the end of the age.

✦ REFLECTION QUESTIONS ✦

What's the specific mission God is calling you to step into next? Be **specific.**

____

____

How can you take bold, mission ready action in the next 24 hours?

____

____

Who can you encourage or invite into their own mission journey?

____

____

✦ FAITH AND ACTION ✦

CONSIDER THIS YOUR mission for today: To surrender to God. Let God guide you. And let God guide your loved ones!

Here's the reality: God loves you and them. God guides you and them. So let the Creator do that! If you feel that fear, ask God to lift it so you can surrender yourself and others to His magnificent love and power.

At the end of the day, write in your Notes to God how you did. Did you surrender everything and everyone to God?

✦ NOTES TO GOD ✦

CREATIVE SPARK

(Week 5 Close)

CREATE FORWARD

(as seen through the lens of the creative spark)

Passage: 1 Timothy 4:14

14Do not neglect the gift that is in you..."

In every mission, there comes a moment where theory ends, and action begins. For a rocket, it's the ignition. For an artist, it's the first stroke on a blank canvas. For a disciple, it's the moment you say: Yes. I'll go.

You've studied the patterns. You've felt the rhythms. You've seen the beauty in the numbers and the wonder in the words. Now comes your invitation: Create forward.

Don't just consume beauty; become part of it. Don't just observe the system; help shape it. Don't just reflect on the spark; carry it into the world.

Creativity is never meant to stay confined.

- A melody is meant to be sung.
- A story is meant to be shared.

- A design is meant to be built.
- A calling is meant to be lived.

In 1 Timothy 4:14, Paul urges the young leader: "Do not neglect the gift that is in you." That word "gift" in Greek is **charisma**, a grace-infused ability from God. You have one.

Maybe it's writing.

Maybe it's organizing.

Maybe it's dancing or designing or dreaming systems that heal injustice.

Maybe it's raising children, building nonprofits, or building bridges, literal or spiritual. Whatever your gift is, it was never meant to sit in silence.

Create forward.

Take what God has stirred in you over these 31+ days, through science and scripture, through rhythm and revelation, and move it into the world.

With intention.

With integrity.

With imagination.

✦ FAITH FORMULA ✦

Creation is the continuation of calling. When you create forward, you participate in God's ongoing work of redemption.

✦ GOD'S CONSTANT ✦

He didn't just create you; He entrusted you to create. And He walks with you into every act of obedience and artistry.

✦ REFLECTION QUESTIONS ✦

What creative expression or calling have you been neglecting out of fear, doubt, or distraction? Be specific.

____

____

What insights or images from the last five weeks keep stirring in your spirit, or how has this journey reawakened something imaginative or purposeful in you?

____

____

____

What fear or friction is delaying your personal launch?

____

____

____

Where do you feel God nudging you to create forward this week in art, in work, in relationships, in worship, or in your witness?

____

____

____

✦ FAITH AND ACTION ✦

- Sketch your calling. Use words, images, or equations to visualize what obedience looks like in this season.
- Make one intentional move this week. Apply for the role, reach out to the collaborator, start the project, write the first sentence, sketch the first design.
- Ask God for boldness. Step into your design and flow with grace. He built you for both launch and legacy.

✦ NOTES TO GOD ✦

CONCLUSION

Faith in Motion. Creation in Progress.

You've reached the end of this devotional, but you've only just launched. Over these 31 days, plus the creative sparks, you've explored a God who speaks through **physics and poetry**, who reveals truth in **equations and parables**, who designs with **precision** and **paints with beauty**.

You've seen that:

- Structure is not separate from spirit.
- Creativity is not a side note to calling, it's central.
- And God doesn't draw the line between science and faith. **He draws the connection.**

Now what?

Now you create forward.

Not just because the world needs better engineers, artists, teachers, coders, leaders, and thinkers. But because the world needs people whose **faith is in motion.**

People who move with both truth *and* tenderness.

People who build systems *and* tell stories.

People who solve problems *and* imagine new possibilities.

People who walk with a God big enough for black holes… *and* near enough for broken hearts.

This is your invitation:

To live like your design matters.

To see the world not only as it is, but as it could be.

To reflect the God who still creates through people like you.

The **formulas** are in you.

The **rhythm** is in you.

The **image of God** is in you.

Now rise.

Move.

Make.

Multiply.

You are mission ready.

You are creatively called.

And the kingdom is waiting for what only you can create.

GLOSSARY OF STEAM TERMS

SCIENCE & ENGINEERING

- **Asymptote** – A line that a curve approaches but never touches. Often used to describe things that draw infinitely close but are never fully grasped, like the mystery of God.
- **Entropy** – A measure of disorder in a system. Over time, **systems** tend to move from order to chaos unless energy is added, a powerful metaphor for spiritual drift.
- **Escape Velocity** – The speed required to break free from a gravitational pull. Spiritually, it represents the force of grace needed to overcome the weight of sin or fear.
- **Feedback Loop** – A system where outputs are fed back in as inputs for growth or correction. Like how conviction or wisdom guides spiritual course correction.
- **Gravity Well** – A region where gravity is so strong that it becomes hard to escape. Used metaphorically to describe habits, emotions, or systems that pull us downward.
- **Load-Bearing Structure** – A part of a system that carries significant weight. Often used to reflect people or

practices that hold spiritual, emotional, or communal weight.

- **Redundancy (Engineering)** – The intentional design of backup systems to prevent failure. Spiritually, it reflects God's provision even when Plan A falls apart.
- **Stress Test** – A way to determine how much pressure a system can endure. Similar to how trials reveal the strength of our faith.
- **System Integration** – Combining subsystems into one functioning whole. Used to describe spiritual alignment, bringing all parts of your life under the lordship of Christ.
- **Thrust** – The force that propels a rocket forward. Represents spiritual momentum or the energy of obedience and grace in action.
- **T-minus** – Countdown language used before launch. Represents spiritual urgency and the readiness to step into your calling.

TECHNOLOGY & DIGITAL LIFE

- **Algorithm** – A step-by-step instruction set used in technology. Spiritually, it symbolizes how God's guidance can lead us step by step, even when we don't see the outcome.

- **Artificial Intelligence (AI)** – Machines that simulate human thinking. Used to contrast artificial systems with the eternal wisdom and presence of God.
- **Binary Code** – The language of 1s and 0s behind all digital systems. A metaphor for simplicity, identity, and the idea that even the smallest combinations can build complex realities.

MATHEMATICS

- **Exponential Growth** – A pattern where small things grow rapidly and multiply. Reflects the nature of Kingdom influence, faith, and obedience.
- **Limit (Mathematics)** – The value a function approaches as it nears a certain point. Often used as a metaphor for the infinite nature of God, approachable but never exhausted.
- **Pi (π)** – A mathematical constant representing the ratio of a circle's circumference to its diameter. It never ends and never repeats, a symbol of God's infinite nature.

THE ARTS & IMAGINATION

- **Creative Spark** – The divine impulse to make, express, or imagine. It reflects the image of God in us as co-creators, storytellers, and beauty-bearers.
- **Design Thinking** – A human-centered approach to solving problems creatively. In a spiritual context, it mirrors how God engineers redemption through both logic and love.
- **Pattern** – Repetition with meaning. Seen in music, poetry, architecture, and Scripture, pointing to a God who reveals truth not just through facts, but through rhythm.
- **Perspective** – The way we see and interpret things. Artistic and spiritual growth often depend on learning to see from a higher, more compassionate vantage point.
- **Rhythm** – The flow of movement, time, or sound. It represents the divine pulse in all of creation, from breath to seasons to worship.
- **Storytelling** – The act of conveying truth through narrative. A sacred tool used by Jesus (parables) and by believers to share meaning, history, and hope.

BONUS

MARVEL, THE MULTIVERSE, AND THE MAJESTY OF GOD:

What comic books can teach us about calling, choices, and the mystery of God's plan

If you've ever watched a Marvel movie, or fallen into a comic book rabbit hole, you've probably heard of the **multiverse**: an infinite web of alternate realities. In one world, Peter Parker is Spider-Man. In another, Gwen Stacy is. In one timeline, Tony Stark survives. In another, he doesn't.

The idea? Every choice leads to a new branch.

Every possibility exists.

Every version of you is playing out somewhere in time.

Fascinating stuff. And while the multiverse may be a work of fiction, it reflects something deeply real: **our hunger to make sense of choices, time, destiny, and calling.**

THE MARVEL OF FREE WILL AND SOVEREIGNTY

In *Doctor Strange in the Multiverse of Madness*, we're reminded that one small decision can lead to a completely different future. In many ways, that's true for us too. One conversation. One act of courage. One prayer. One yes.

But here's the difference:

In Marvel, the multiverse is chaotic. In God's Kingdom, the Author of Time is in control. Where the Marvel universe bends under the weight of infinite what-ifs, **God holds every possible path in perfect knowledge**. He doesn't get overwhelmed by your choices. He already sees them and **redeems them.**

He doesn't just know one version of your future. He knows every possible fork in the road, and He still chooses to walk with you. Not to trap you in one "correct" timeline, but to offer you **constant grace, wherever you are.**

NEXUS MOMENTS AND KINGDOM MOMENTS

In the Marvel world, a **nexus event** is a moment that branches off reality into something entirely new. In God's world? That's called **a calling**.

It's that moment when everything shifts:

- The whisper to forgive.
- The divine opportunity to lead.
- The choice to surrender your life to Jesus.
- The moment you say yes, and heaven moves.

You don't need a multiversal portal to access your potential. You just need to listen for God's voice and respond.

MULTIVERSE MYSTERY. ETERNAL SECURITY.

The beauty of the Marvel multiverse is the imagination. The beauty of God's sovereignty is **security**.

You don't have to live in fear of making the wrong move.

You don't have to wonder if you missed the "real" version of your life.

You don't need to fix the timeline. Jesus already did that on a cross.

So be faithful where you are.

Live boldly.

Love fully.

Act justly.

Launch deeply.

Because while Marvel imagines endless possibilities, **God promises one glorious ending**:

Every tear wiped away. Every knee bowed. A new heaven and a new earth. And you, with Him, forever.

✦ FINAL THOUGHT ✦

You're not a variant. You're **chosen**.

You're not lost in time. You're **anchored in eternity**.

You don't need infinite timelines. You need **faith in the One who never changes**.

Suit up. You are mission ready. And the Kingdom needs you, right here, in *this* reality.

ABOUT THE AUTHOR

CHRISTOPHER M. JONES, Ph.D. is one of the few people who can talk as fluently about Biblical theology and Christian principles as he can about quantum mechanical states and differential equations. He is among an even smaller set of people who can do that while recalling stories of fishing, playing sports, organizing in urban and rural communities, being a military spouse and doing his girl dad thing. He brings all of this to bear in Quantum Pearls.

A native of Pine Bluff, Arkansas, Dr. Jones was the first African American nominee for Governor by a major party in Arkansas. His love for science took him to Morehouse College, where he earned physics and math degrees, to NASA where he interned under an astronaut and worked on a plasma rocket ship, and to MIT where he earned two master's degrees—nuclear engineering and technology and policy—and a Ph.D. in Urban planning. The son of two preachers, his love for Christ drove him to teach

Sunday School at a young age, to lead the College ministry at Morehouse, to jump start the church's young professional's ministry while at MIT, and to become an ordained minister. His love for community and people opened the door for him to teach high school math, serve as an Assistant Dean for Graduate Education at MIT, run a community-based organization in Boston (The Dudley Street Neighborhood Initiative), lead a maker space in Arkansas (The Arkansas Regional Innovation Hub) and eventually run for Governor of Arkansas. The experiences and tools that Dr. Jones has learned along the way have helped him daily excavate his own life and find valuable quantum pearls that he shares in this devotional. He is the author of *Quantum Pearls: Finding Spiritual Wisdom in the Mundane Moments*, which he calls QP1. More than anything, Dr. Jones loves his wife of 20 years, Dr. Jerrilyn Jones, who is an emergency room physician and combat veteran of the United States Air Force. Together, they have three beautiful daughters, Jordyn, Janelle, and Jasmine.

www.ingramcontent.com/pod-product-compliance
Ingram Content Group UK Ltd.
Pitfield, Milton Keynes, MK11 3LW, UK
UKHW020123290926
15471UKWH00003B/238

9 798989 090945